Heinz J. Teige

SPACE AGE CRYSTALS®

THE ART AND SCIENCE OF CRYSTAL GROWING ON EARTH – AND IN SPACE

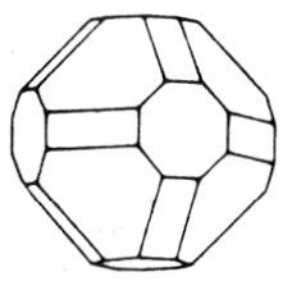

COVER PHOTO: The Space Shuttle CHALLENGER as seen from the unmanned, free-flying satellite SPAS-01 during mission STS-7. Seven GET AWAY SPECIAL containers are mounted in the Shuttle's cargo bay, one of them housing a student crystal growing experiment (see chapter 4.1 and 4.2). In addition, crystals of germanium selenide, a semiconductor, were grown onboard the orbiter.

ISBN 0-9633525-0-4

9 780963 352507 90000

THE KRISTAL CORPORATION

P.O. Box 1677
Sparks, Nevada, 89432
(702) 359-0411
USA

First Edition. Printed in Germany.

International Standard Book Number: 0-9633525-0-4
Library of Congress Catalog Card Number: 92-72974

Special Thanks and Acknowledgements:

National Aeronautics and Space Administration
Dr. C. Bugg, University of Alabama at Birmingham
Space Studies Institute
Edmund D. Burke
GAS # 450 team members
Dr. Brenda Forman
NASA ACTIVITIES
National Space Society
Boeing Aerospace
MR Semicon
Dornier System GMBH

PICTURE CREDIT:

NASA: Front + back cover, Fig. 3, 4, 6, 12, 13, 19, 50, 51, 52, 53, 54
University of Alabama at Birmingham: Fig. 8, 9, 10, 11
GAS # 450 team: Fig. 20, 21, 22
Space Studies Institute: Fig. 7, 14, 15, 16, 17
Boeing Aerospace: Fig. 18
MR Semicon: Fig. 1
Dornier Systems GMBH: Fig. 5

Table of Contents

Fig. 1 Man-made crystals are built into almost every electronic device, and without them our technological civilization could no longer function. This photo shows an automated production facility (a "High Pressure Czochralski Puller") for growing large crystals of gallium arsenide (GaAs) which are used in the semiconductor industry.

1.1 CRYSTALS – WONDERS OF NATURE

The Greeks in antiquity called frost and ice "krystallos" – hence the word "crystal" which was originally given to clear rock-crystals only. Today this term is used for all solid substances whose chemical building blocks – atoms, ions and molecules – are set up three dimensionally in periodical sequence, and thus form a stereoscopic "crystal lattice". During the growth of a crystal as many as 100 000 000 000 000 000 000 (or 10^{20}) atoms are lined up per cubic millimeter (about one fifteenth of a cubic inch) according to a strict geometrically fixed order.
The remarkably symmetrical shape of a crystal and its regular surfaces surrounded by straight edges are an external feature of the crystalline state. But the name "crystal" is also to be given to any quartz that has been rolling around in a river and whose edges and reflecting surfaces are already completely polished. Its inner crystal structure, however, has remained unchanged after these external mechanical events; and even the smallest grain of sand, once a component of solid rocks, and, after their decomposition, crushed and polished during transportation in the rivers and seas through many thousands of years, has the same crystalline structure as a quartz crystal defined by surfaces, and weighing several tons. Such huge crystals are of course extremely rare, since the growth conditions inside the earth's crust normally do not leave enough space for an individual crystal to develop freely. Nevertheless, huge quartz crystals measuring 7 m (about 23 feet) in circumference and weighing 40 tons were found in Madagascar, and from the Black Hills of South Dakota came spodumene crystals of up to 10 m (about 33 feet) in length. Today's record holder is an 18 m (60 feet) long beryl crystal, also from Madagascar, with a weight of 36 tons and a diameter of 3.5 meters (12 feet)!

1.2 THE FOUR STATES OF MATTER

In order to understand the processes which lead to the formation of crystals we need some basic knowledge about atoms, the building blocks of matter. Ninety naturally occuring types have been discovered up to now. They are so small that it would take, for instance, 20 million hydrogen atoms to form a line of just one millimeter (or 500 million for one inch). Due to their minute size our senses are unable to perceive them directly, and though atoms had already been postulated in ancient Greece it took another 2000 years and the advent of modern science to experimentally prove their existence.
All substances and objects surrounding us are made of atoms which tend to join and thus form "molecules". Water (H_2O), for instance, consists of two hydrogen atoms (H) and one oxygen atom (O). It occurs in three different forms as vapor, fluid or as ice and snow. What is the difference between these three "states of matter"? We know that each of them is associated with a certain temperature range. But what is "temperature"?
Even in a solid object atoms and molecules oscillate constantly, and the higher the temperature the faster they move. And vice versa: the cooler an object, the slower the movement of its atoms. Only at the deepest possible temperature ("absolute zero", -273.15° Celsius or -459.67° Fahrenheit) atoms are almost completely at rest.
In the gaseous state of matter atoms and molecules move independently in straight lines. If they collide with each other or with other objects they bounce off like billiard balls. Lowering the temperature of the gas results in a decrease of its "kinetic energy" (energy of movement) so that collisions become less intense. A further decline in temperature will finally lead to a point where the gas "condenses": molecules no longer bounce off but remain loosely attached to each other. In case of water this condensation point (or boiling point) is reached at 100° C (212° F). For oxygen it is -182.97° C (-297.35° F) and for gold 2960° C (5360° F). The gas has now become a fluid where the average distance between atoms or molecules is much smaller than before. They still move, however, quite intensely and neighboring molecules are continuously regrouped.
After a further decrease in temperature the freezing point (or melting point) will finally be reached (water 0° C or 32° F, oxygen -218.7° C or 361.7° F, gold 1063° C or 1945.4° F). The kinetic energy of the atoms and molecules is now so small that strong and permanent bonds become possible: a solid forms, and in most cases it is a crystal whose building blocks are lined up in a periodical sequence and thus form a crystal lattice. There are, however, a few "amorphous" substances like glass, soap or plastics which are solid but whose components are not arranged in a firm geometrical order and which therefore do not form crystals.
Compared to fluids or even gases the movement of atoms and molecules inside solids is drastically re-

duced to more or less intense oscillations around their given place in the crystal lattice. As a result a solid contains between 1,000 and 100,000 times as many atoms as a gas of the same volume.
Besides the three states of matter mentioned before there is a fourth one called "plasma". As we have seen it is an increase in the kinetic energy of its atoms which transforms a solid first into a fluid and then into a gas. If you now continue to heat the gas its atoms get faster and faster and the intensity of their collisions steadily increases. This finally leads to a point where the atoms "damage" each other – they get "ionized" which means that they are broken down into their electrically charged building blocks and thus form a plasma.
The sun (whose interior has a temperature of between 15 to 20 million degrees Celsius) and every other star consist of plasma, and since the total mass of all solar planets is comparatively small (it would take the mass of 333,434 earthlike planets to equal that of the sun) the majority of the presently known mass of the universe therefore exists in the fourth state of matter.

1.3 MINERALOGICAL TREASURE CHAMBERS

As we have seen the formation of crystals is the rule whenever fluids are sufficiently cooled down to become solids. Up to now more than 20,000 crystallized substances are known, and new ones are discovered regularly. About 3,000 of those can be found in the earth's crust and are called "minerals". Most are quite rare, and just a few (about 40 to 50) are relatively common like quartz, feldspar or mica. These constitute the major components of the various types of rocks and are therefore called "rock-forming minerals".
The reason for the formation of all earthly crystals, minerals and rocks is the temperature difference between the earth's core (at least 5,000° C) and space (near absolute zero). As we know from deep mines and drilling operations the temperature of the earth's crust shows an increase of 1° C (1.7° F) every 33 meters (100 feet), and only 50 kilometers (31 miles) below the surface of the continents all rocks have become liquified. In the deep sea regions the thickness of the solid crust is often reduced to just 4–5 km (2.5–3 miles). If we could shrink the earth (diameter 12,757 km or 7,927 miles) to the size of a globe (30 cm or about one foot) the solid continental crust would measure one millimeter (1/25th of an inch). And by using the same scale earth's mantle of air would just equal the thickness of a piece of paper! In the transition zone between the solid crust and the liquid magma temperatures range from 900° to 1,200° C (1,600° to 2,200° F) and the pressure is several thousand times higher than on the earth's surface. Magma rises from great depth, cools down and sinks back. This process is called "convection" and you can observe its basic mechanism in another context during crystal growing experiments.
The continuous movement of magma has of course far reaching consequences for the earth's surface. Whole continents drift, the crust breaks and is pushed up to form mountains. Earthquakes accompany these events, and magma reaches the surface during volcanic eruptions. It consists of molten rock which contains about 75% silicon and oxygen and is saturated with various gases and water vapor. The simplest chemical combination of silicon and oxygen is silicon dioxide (SiO_2). If magma rises in the fractures of the crust it cools down. The pressure decreases simultaneously which leads to the formation of bubbles like in a just opened bottle of carbonated mineral water. A further drop in temperature will turn the magma into a solid, and the bubbles will form cavities. Water vapor condenses there and SiO_2 is dissolved out of the surrounding rock.
As we know from our everyday life the solubility of some substances (like sugar) in hot water is greater than in cold water. The same principle applies to our SiO_2 solution: in the course of many millennia it continues to cool down. At some point it will contain more dissolved SiO_2 than possible at this lower temperature, and the excess silicon dioxide now has to go out of solution into the solid state of matter. Quartz crystals start to form on the walls of the cavity. Depending on minute concentrations of other substances in the solution this quartz can either be clear rock-crystal, purple amethyst or brown smoky-quartz. If the temperature continues to decrease the solution will steadily deliver more building material for the crystals which now grow deeper into the cavity. The crystals are, however, still far below the earth's surface and it takes additional geological forces to get them there. Magmatic convection currents might push up the earth's crust in this area during the formation of mountains, and wind and weather will gradually erode the higher layers of rock. In high mountain regions, for instance, the forces of erosion need 2,000 years to carry one meter (3.3 feet) of rock away. Sooner or later the crystals will be at the earth's surface, and perhaps a lucky mineral collector will find these treasures of nature before the forces of erosion grind them down to dust and sand.

2.1 MAN-MADE RUBIES AND DIAMONDS

Ever since antiquity it has been known that solutions of various salts produce big crystals when they evaporate. Aristotle (384-322 B. C.), for instance, describes the production of table salt ($NaCl$) by crystallization. But it was not before the end of the nineteenth century that the alchemists' dream of creating gemstones became finally true. In 1891 the French chemist Verneuil developed a method of growing large ruby crystals by melting aluminium oxide (Al_2O_3) and small amounts of chromium oxide at 2,000° C in a flame of burning hydrogen and oxygen – the same chemical reaction, by the way, which is used in modern rocket engines to transport our astronauts into space! Powdered aluminium oxide falls through this hot flame and forms molten droplets which precipitate onto the surface of a tiny seed crystal and thus allow it to grow. Verneuil's technique was so ingenious that it is still used today for our present annual production of about 250 tons of ruby. Due to its hardness it is not only used in jewelry but has numerous technical applications, for instance as bearing stones in clocks.

The production of diamonds is much more difficult than the synthesis of ruby. It is possible today to grow flawless specimens weighing more than one carat by using high temperatures (up to 5,000° C) and extremely high pressures (up to 100,000 times higher than our normal atmospheric pressure), but presently these techniques are still so complicated that good synthetic diamonds cost more than natural ones. The chemical composition of a diamond is exactly the same as that of ordinary coal or graphite: it is just pure carbon. The difference lies in the denser crystal lattice of diamonds which were formed at great depths (150–200 km or 100-120 miles) under high pressures and high temperatures and then transported towards the earth's surface through volcanic "pipes". Today an average of about 3.5 tons of magmatic rock has to be processed in order to obtain just one carat of diamond.

Since it is the hardest known mineral (hardness 10 on the Mohs' scale) diamond has numerous technical applications. Presently the industrial cultivation of synthetic diamonds mainly aims at this market. You might well have some small man-made diamonds around your house since they are also used for saw blades, drills and grinding wheels.

2.2 CRYSTALS AND TECHNOLOGY

In recent decades man-made crystals have become so important for our life in a modern technologically advanced society that without them civilization as we know it today would no longer be possible. Yet few people are informed about their usefulness or of the vast industry that creates them, although these crystals are built into almost every electronic or optical device made today. Without the ability to grow crystals of such esoteric substances like silicon, germanium or YIG, the technological progress seen in recent decades would not have been possible. And it is also evident that future developments in electronics and optics will depend on the successes of crystal growers.

Products that we take for granted today like computers, TV sets, radios, telephones or your quartz watch contain "solid state" components – small chips of single crystals with a complex fabrication history. They also help to regulate the power supplies of our cities, manage our bank accounts and credit card purchase records, direct airline traffic and reservations, help diagnose diseases, store information, and – through solar cells – may ultimately provide the key to the production of limitless energy.

The earliest radios used crystals of galena (lead sulfide), and operators had to move a wire on the surface of the crystal to find the best spot for reception. By 1947 semiconductor crystals made out of silicon and germanium lead to the first transistor, and a continuous improvement of the crystals enabled technicians to develop integrated circuits and microchips.

Due to an ever increasing integration density – more and more electronic components are put together in a given area – requirements for the crystals' quality and purity are steadily being raised. Impurities of one part in a trillion can change the semiconductor's performance (for better or worse!), and today we have reached a point where even minute gravity-related disturbances during the crystals' growth seriously influence their effectiveness.

Man-made crystals are also used in lasers – the first one, for instance, which was constructed in 1960, contained a synthetic rod-shaped ruby crystal, and emitted an intense beam of red light. A laser is an extremely versatile instrument which has a wide variety of applications ranging from surgery to arms technology and telecommunications. For the latter use the laser light has to be modulated – for instance turned on and off like a flashlight to transmit a coded message. The message should of course be sent as fast as possible, and instead of turning the laser itself on and off another type of man-made crystal is

placed in the path of the beam through which the light has to pass on its way to the receiver. The crystal used for this purpose has a very interesting characteristic: it changes its refraction power in an electric field which means that the beam is bent away from the receiver if the electric field is activated. Therefore, by turning the field on and off, the receiver sees the laser light go on and off in the same rhythm. This way you can transmit messages as with an ordinary lamp, only at the speed of light!
The synthetic crystals used for this purpose are called "electro-optic light modulators" since they change their optical characteristics in response to an electric field. Some of these crystals are grown from an aqueous solution – with basically the same simple procedure that is used in the SPACE AGE CRYSTALS® series of crystals growing kits.

2.3 HOW TO GROW SPACE AGE CRYSTALS®

Crystals have been growing inside the earth's rocky crust for billions of years, and they are still growing there at this very moment – miles beneath your feet!
Wouldn't it be fascinating to observe this with your own eyes? But unfortunately a natural quartz crystal, for instance, takes thousands or even millions of years to grow, and obviously none of us has that much time.
The SPACE AGE CRYSTALS® series of crystal growing kits, however, reduces this space of time to just a few days. Simple experiments allow you to observe for yourself how sparkling crystals (see Figures 32–49) form seemingly out of nowhere and magically acquire their symmetrical shapes and glittering surfaces.
Magic, however, is not exactly the right word to describe how these crystals form. Be it inside the earth's crust or in a crystal growing vessel, it's the forces of nature that build crystals out of atoms and molecules.
A crystal with a volume of one cubic centimeter (about 1/15 of a cubic inch) contains 10^{23} atoms – that's a one with 23 zeros – all of which were built into the crystal by the invisible construction crews that we call "the forces of nature".
And these construction crews really do a magnificent job:
Imagine yourself with a huge pile of LEGO blocks – again, a one with 23 zeros – and you start putting these blocks together. You are fast, and add one block every second. And you have help. Each and every human being on this planet assists you, and all of you work for 24 hours a day – without any sleep or meals! How long do you think this would take? Well, all of mankind would be busy for about ONE MILLION YEARS!
The invisible construction crews that work inside a crystal growing vessel do the same job within just a few days. And although our senses are too imperfect to observe this directly we can certainly judge the results: Sparkling, multi-colored crystal clusters with a diameter of up to 10 inches that rival any "natural" mineral specimen!
You do not need any special equipment to grow these crystals – just some ordinary household items like plastic vessels or jars, and rocks of various sizes. Although the exact crystal growing recipes depend on the kind of kit you are using, the basic techniques are the same (see Figures 2 a–f).
The crystal growing substances which are contained in the kits are dissolved in hot water. You then place a rock into your crystal growing vessel and allow the solution to cool down. At room temperature the solution contains more dissolved chemical than it can hold, and these excess molecules cannot stay dissolved. They have to change over into the solid state of matter and start to form crystals.
Initially, these crystals grow quickly, and due to this fast growth crystal defects occur during the early phases of the crystals' construction. Small amounts of the solution are built into the crystals which results in a strong clouding. But with every molecule which leaves the solution its "supersaturation" decreases. The growing rate slows down which in turn leads to less crystal defects. And within hours the newly formed parts of the crystals become brilliant and clear.
After a few days the solution is saturated which means that it contains just as much chemical as can be dissolved at room temperature. The growth through "supercooling" (the lowering of the temperature) is now completed, and you can remove the crystals from the crystal growing vessel.
If, however, you leave them in the solution they will continue to grow – although for different reasons. Slowly but steadily water will evaporate from the solution. The molecules of the crystal growing substances which had been dissolved in this water also have to leave the solution, and become attached to the crystals.

How to grow SPACE AGE CRYSTALS®

Growing SPACE AGE CRYSTALS® is easy – and it's fun! Details of the crystal growing procedures depend on the type of kit you are using, but the basic recipe is the same:

Fig. 2a Dissolve the crystal growing substance in hot water.

Fig. 2b Select a rock, and place it into a crystal growing vessel.

Fig. 2c Add the solution, and let it cool down to room temperature.

Fig. 2d After just a few hours the first crystals will start to grow.

Fig. 2e They will continue their growth for several days ...

Fig. 2f ... and can be removed from the solution after about a week.

This is of course a fairly slow process, and depending on the humidity and temperature in your area it might take weeks. But due to the slow growth there will only be very few defects and the crystals will become extremely clear.
Growing SPACE AGE CRYSTALS® is a unique way to create your own mineral collection, and can actually become a fascinating hobby. In addition, it serves as an introduction to the science of crystal growing. With SPACE AGE CRYSTALS® kit no. 6434 ("Clear Column"), for instance, you can actually duplicate experiments that astronauts conducted onboard SKYLAB, our first space station.
THE KRISTAL CORPORATION, manufacturer of the series of SPACE AGE CRYSTALS® kits, also offers a two-hour video under the same title. It includes step-by-step instructions for the basic experiments, as well as time-lapse sequences of growing crystals. In addition, you can observe how scientists grow crystals on earth and in space. This video also includes interviews with students who conduct crystal growing experiments onboard the Space Shuttle (see chapters 4.1 and 4.2).

3.1 CRYSTALS – "MADE IN SPACE"

Among the first commercial products manufactured aboard earth-orbiting space stations will be high-purity semiconductor crystals. Our rapidly advancing electronics industry has an ever increasing need for single crystals made out of substances like silicon, germanium, gallium arsenide and cadmium telluride. Presently 5,000 tons of such crystals are produced annually with a value of about three billion dollars, and every year this market grows by 20%.
Due to the steady improvement and miniaturization of electronic components crystals have to be perfected on a continuous basis. The number of defects in the crystal lattices must be further minimized, and certain new desirable impurities which improve the crystals' electronic characteristics have to be induced in a controlled way ("doping").
Ideally a specific type of atom would be located at a known place in the crystal lattice. A difficult problem if you keep in mind that there are 10^{20} atoms in just one cubic millimeter of a crystal! But at any rate the flow of material and temperature has to be carefully controlled during the crystallization procedure, and it has turned out that gravity seriously limits the amount of control we can exercise on a growing crystal. The reason for this is convection – basically the same phenomenon that causes the movement of magma inside the earth (see chapter 1.3).
We all know convective forces from our everyday life. A glider pilot, for instance, uses them in the form of upcurrents to reach greater heights. Air near the ground is heated, expands and thus gets a lower specific weight than the cool air above it. Like air bubbles in water warm air is subject to buoyancy and will therefore rise.
SPACE AGE CRYSTALS® experiments allow you to observe these convective forces for yourself: since molecules leave the solution and are integrated into the growing crystal, the specific weight of the solution in the immediate vicinity of the crystal changes. Convective currents move towards the surface.

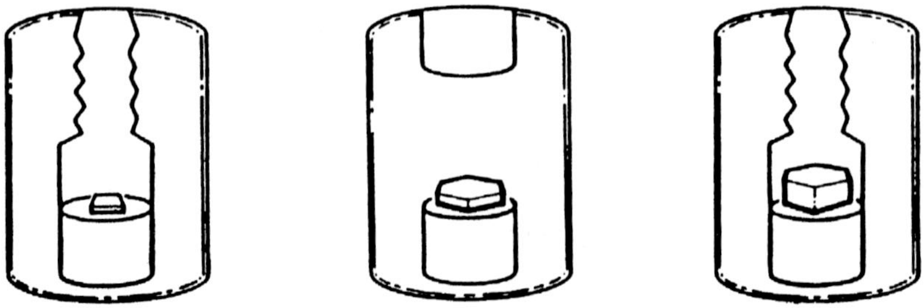

Fig. 3a Fig. 3b Fig. 3c

SHUTTLE EXPERIMENT: GROWTH FROM A SOLUTION

The TGS solution is heated while the protective cap remains on the seed crystal (Fig. 3 a). As soon as the cap is raised the crystal starts to grow (Fig. 3 b). 24 – 48 hours later the crystal's growth is completed and the cap is replaced (Fig. 3 c).

Uncontrollable turbulences in these currents influence the flow of material and the temperature which results in an increase in crystal defects.
In a weightless environment (or rather under "microgravity", since even in space stations there are residual acceleration forces in the order of 1/10,000 g) this "gravity driven convection" is negligible which results in a substantial improvement in the size and quality of the crystal.
Today crystal growing experiments have become an important part of nearly every manned space flight with scientific objectives. The crew of STS-61A (Spacelab D-1 mission, November 1985), for instance, conducted a total of 75 experiments, and 13 of these were designed to improve various techniques of producing single crystals from a solution, out of a melt or from the vapor phase ("sublimation"). The technical equipment for these experiments is of course quite sophisticated, but the basic physical mechanisms and procedures are the same as in the SPACE AGE CRYSTALS® experiments. During the Shuttle mission STS 51-B ("Spacelab 3", April 1985) payload specialist Dr. van den Berg – who, by the way, was selected for that mission because of his expertise in crystal growing – grew TGS crystals [triglycine sulfate, $(NH_2CH_2COOH)_3 \cdot H_2SO_4$], from an aqueous solution. This substance is an important infrared detector material, which in contrast to other types of detectors does not require expensive and elaborate cooling systems. Triglycine sulfate crystals grown on earth have microscopic defects caused by gravitationally induced convective currents, which reduce their electrical and optical performance. Better crystals could have a significant technological impact by improving, for instance, earth observation cameras, telescopes or satellite detection systems.
This "Fluid Experiment System" included several crystal growing cells, an elaborate optical monitoring assembly, a control panel, and associated electrical hardware. Before launch the cells were filled with about one liter of high-purity crystal growing solution (see Fig. 3). Seed crystals with a diameter of 1.0 to 1.5 cm (about 0.5 inches) were attached to "cold fingers" – insulated stings which extract heat from the crystal – and covered with a cap to prevent a premature contact between the seed and the solution which was then heated to 70° C (158° F). After all TGS had been dissolved the temperature of the solution was lowered to 42° C (108° F) and the protective cap removed from the seed. Through the cold finger heat was now extracted from the crystal in a carefully controlled manner which resulted in the crystal's slow but uniform growth of about one millimeter per day.
After the return of the shuttle NASA's official mission report described the preliminary results of this experiment:
"During the first flight of this new FES facility, two TGS crystals were successfully grown in space from a liquid. The solution growth process was recorded by video imaging and by a laser holographic system that will render three dimensional images. For the first time, scientists were able to see in detail

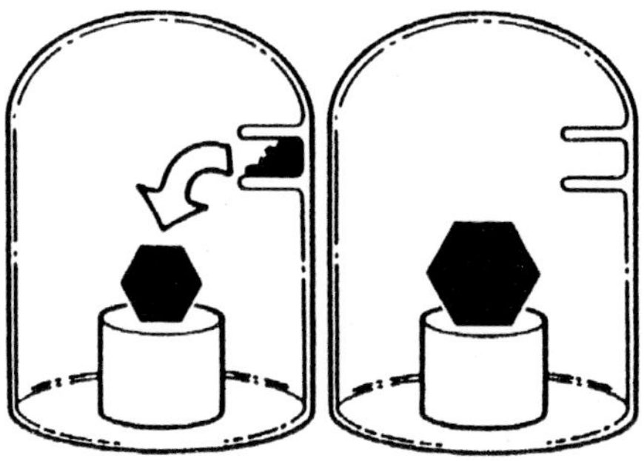

Fig. 4a Fig. 4b

SHUTTLE EXPERIMENT: GROWTH FROM A VAPOR
The source material is heated and sublimes (Fig. 4 a). It spreads through the ampoule and is integrated into the growing seed crystal (Fig. 4 b).

the crystal growth process and determine the difference between crystal growth on the ground and growth in microgravity where convection effects are negligible. Visual observations by the crew provided invaluable real-time description of the crystal and aided investigators on the ground as they made decisions to control the progress of the investigation." Another interesting experiment of mission 51-B was the growth of mercury iodide (HgI_2) crystals which have considerable practical importance as sensitive X-ray and gamma-ray detectors. Presently available detectors must be cooled to near liquid nitrogen temperatures, and because bulky cryogenic cooling systems are not needed, mercury iodide crystals

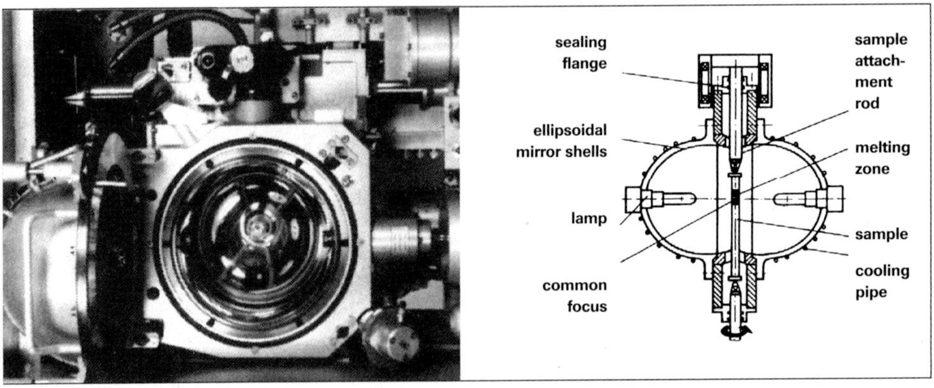

Fig. 5 Semiconductor crystals are grown in this "Mirror Heating Facility" by melting a rotating silicon rod and transforming it into a single crystal.

could be useful in portable detectors for nuclear power plant monitoring, natural resource prospecting, biomedical applications, nuclear weapons work, and in astronomical instruments. As with TGS crystals, however, unstable convective flows in ground-based processing cause defects in the finished crystals. In addition, mercury iodide has a fragile structure that can be deformed during growth by the stress of the crystal's own weight.

Although this substance is crystallized from a vapor the basic physical mechanisms are similar to the previous experiment:

The source material (HgI_2) and a seed crystal were enclosed in a sealed ampoule inside a furnace. When the ampoule was heated to 100° – 110° C (212° – 230° F) the source material "sublimed" (went directly from the solid state of matter into the gas phase without passing through the liquid state). It spread through the ampoule and was integrated into the growing mercury iodide crystal seed which had been attached to a cold finger with a temperature of 40° C (104° F).

NASA's official mission report summoned up the experiment:

"A mercury iodide crystal measuring 14 mm x 8 mm x 7 mm, about the size of a sugar cube, was successfully grown from a seed crystal 20 times smaller by a vapor transport process. The crystal was grown at a carefully controlled rate that varied from 1 mm to 3 mm per day over a 104-hour growth period. Downlink video and crew reports of visual observations were major benefits to investigators on the ground, who were able to follow the course of the investigation in detail."

Semiconductor crystals have been a prime investigation target ever since the days of Skylab. Fig. 5 shows a special furnace, the "Mirror Heating Facility" which was flown on various missions like Spacelab 1 and Spacelab D-1. It reaches temperatures of up to 2,100° C and was used to grow silicon crystals by "zone crystallization". Two spherical mirrors concentrate the energy of two halogen lamps in a common focus where a small part of a rotating polycrystalline silicon rod is melted. As the melt cools down it forms a single crystal, and by moving the focus, an adjacent area of the rod is melted into which the single crystal can continue to grow. Within a matter of hours the whole rod is transformed into a single silicon crystal.

Doped silicon is the most important material for manufacturing integrated circuits. Crystals are grown on an industrial scale by zone crystallization and by the Czochralski process (pulling from the melt). In all crystals grown on earth variations in dopant concentration are found. The ensuing local variations of the electrical properties become problematic due to the continuing miniaturization of electronic components. Variations in dopant concentration result in parallel "striations" which can be revealed by etching techniques. Striations are the consequence of fluctuations in temperature and concentration, and these fluctuations in turn result from turbulent flows in the melt which are partially caused by gravity-driven convection.

Besides their numerous technical applications space grown crystals are also of great importance for medicine and pharmacology. These disciplines require well-shaped and sufficiently large single crystals of organic macromolecules, especially proteins, in order to analyze their molecular structure by apply-

ing X-ray crystallography. The growth of such crystals in earthbound laboratories is disturbed by convection and turbulences due to different densities of the solutions. Spacelab 1 experiments proved that by using the same techniques in microgravity the crystals' size and growth rate could be increased significantly: crystals of the enzyme β-galactosidase (molecular weight 465,000) were 27 times larger than those grown on earth. And for the enzyme lysozyme (molecular weight 14,307) the size could even be increased by a factor of 1,000! The corresponding solution growth techniques can be applied to a wide range of organic substances. They might well result in a better understanding of our metabolic processes and consequently lead to new drugs.
Due to their great importance and extremely promising results these Shuttle-based protein crystal growth experiments will now be described in greater detail.

3.2 A USER'S EYE-VIEW OF THE SPACE SHUTTLE

The following article is reprinted from a NASA publication ("NASA ACTIVITIES", November/December 1990) and contains an interview that NASA conducted with Dr. Charles E. Bugg of the University of Alabama's Center for Macromolecular Crystallography concerning the shuttle borne Protein Crystal Growth Experiments. Dr. Bugg heads the effort to understand protein crystal growth in microgravity conditions.

NASA Activities: What unique capabilities enable these experiments to be successful?

Dr. Bugg: The unique capability is a carefully controlled microgravity environment. Microgravity offers two primary advantages: First, sedimentation processes are eliminated in the absence of gravity. On earth, when crystals grow from solution, they normally sink to the bottom of the container. This sedimentation causes the crystals to grow in a non-uniform environment, resulting in deformed crystals. In the absence of gravity, crystals grow suspended in the solution, bathed on all sides uniformly by the growth medium. Thus, the crystals can grow uniformly in microgravity. Secondly, in the absence of gravity, density-driven convective flow is essentially eliminated. Convective flow results any time a crystal grows from a solution on earth. Convective flow can interfere with controlled uniform growth of protein crystals.

NASA Activities: What is the necessity of the crew to the experiments?

Dr. Bugg: These experiments have been designed to be performed in the mid-deck of the space shuttle. This permits experiments to be performed by members of the crew, and the equipment has been designed to take advantage of crew operations. By using crew members, automation requirements have been minimized. This increases reliability, and permits experiments to be performed where the outcome is somewhat uncertain. Trained crew members can respond to unexpected experimental events, which adds new dimensions to the possibilities of doing forefront research on the space shuttle. We routinely use crew members to activate and deactivate the experiments, to photograph interesting features that result in the crystal growth solutions, and to note any unusual events that occur during the experiments.

NASA Activities: How do we handle the data to maintain accuracy?

Dr. Bugg: There are a number of uncertainties involved in any space experiment. There are a number of other activities occurring on the shuttle, which can interfere with the performance of ideal experiments in protein crystal growth, so control of accuracy in experimental procedures is not always completely ideal. The availability of crew members permits us to note and respond to any unusual occurrences that are unexpected. One way that we try to control accuracy is by performing repetitive experiments, both on an individual shuttle flight and on successive flights. Each protein examined on STS-29 was included in several different experiments on that flight. In addition, most of these protein crystal growth experiments either have been done on earlier flights or will be repeated on later flights.
We try to obtain detailed photographic records and continuous recording of key parameters, such as temperature, in order to document the experimental results in orbit. The samples are returned immediately to Birmingham for complete laboratory analysis within several hours of the time that the space shuttle lands, and these detailed analyses allow us to analyze the accuracy of the data.

NASA Activities: What is the quality of data and experimental results due to use of the shuttle?

Dr. Bugg: The main advantage of the shuttle is that it provides a microgravity environment for performing protein crystal growth experiments. There are some significant advantages of the shuttle, relative to other options for accessing space. The shuttle permits us to load our sensitive biological materials immediately prior to launch, and to retrieve these materials from the mid-deck immediately after the shuttle returns to earth. The shuttle has a mild re-entry, with low g-levels, which permits us to return crystals to earth without significant physical damage. The shuttle provides crew members who are able to perform complex experiments.

The main disadvantage of the shuttle is that it is in orbit for only short periods of time, which limits the number of proteins that can be examined. Another disadvantage of the shuttle is that disturbance due to crew activities, such as exercise regiments, can interfere with certain types of crystal growth experiments.

Overall, the experimental results to date indicate that the space shuttle can be used in certain cases to produce protein crystals that are significantly better than the best crystals of these same proteins produced on Earth. These higher quality crystals permit unique biomedical and pharmaceutical development projects to proceed.

NASA Activities: What are the future plans for Shuttle in your research?

Dr. Bugg: We are scheduled to fly on a number of shuttle flight missions over the next six years. These flights will permit development of new approaches for protein crystal growth in space, and the development of new hardware for optimizing the effects of microgravity on protein crystal growth processes. We now have 60 Co-Investigators from academic, government and industrial laboratories working with us on this project. These various Co-Investigators are studying a wide range of different proteins and different crystallization problems that are likely to benefit from microgravity research in this area. Future plans include the development of equipment and techniques that permit dynamic control of protein crystal growth processes in microgravity.

NASA Activities: What are the medical advantages of the research?

Dr. Bugg: Structural biology is of major basic interest in all fields of medical research. Crystallography is the only technique for determining the arrangement of atoms in the three-dimensional structures of proteins and other large biological molecules. The structural information gained by studying crystals of biological materials permits investigators to understand how various important molecules function in biological systems, and how these functions can be modified in order to solve different specific medical problems.

From the applied standpoint, the information gained by protein crystallography is extremely useful in designing new pharmaceuticals for treating a wide range of human diseases.

Experiments to be included in future shuttle flights are directed at proteins and other biological molecules that are of key importance in developing pharmaceuticals for use in cancer, AIDS, arthritis, diabetes, influenza, cardiovascular diseases, emphysema, and in a variety of other illnesses.

The overriding limiting step in applications of protein crystallography is protein crystal growth. It is not possible to pursue these structural studies, or to use the structural information for designing new pharmaceuticals, unless highly-ordered crystals of the materials can be obtained. Crystallizing most of these biological macromolecules is extremely difficult on earth. Results to date indicate that microgravity can be used effectively to greatly enhance the quality of biological crystals, and it is expected that the future space experiments will further enhance the potential of microgravity protein crystal growth.

3.3 PROTEIN CRYSTAL GROWTH ON STS-31

For each Shuttle mission NASA prepares a detailed Press Kit for distribution to the media which describes the individual mission, its objectives and experiments. The following article is reprinted from the SPACE SHUTTLE MISSION STS-31 PRESS KIT.

The Protein Crystal Growth (PCG) payload aboard STS-31 is a continuing series of experiments leading toward major benefits in biomedical technology. These experiments are expected to improve food production and lead to innovative new drugs to combat cancer, AIDS, high blood pressure, organ transplant rejection, rheumatoid arthritis and many other medical conditions.

Protein crystals, like inorganic crystals such as quartz, are structured in a regular pattern. With a good crystal, roughly the size of a grain of table salt, scientists are able to study the protein's molecular architecture.

Determining a protein crystal's molecular shape is an essential step in several phases of medical research. Once the three-dimensional structure of a protein is known, it may be possible to design drugs that will either block or enhance the protein's normal function within the body or other organisms. Though crystallographic techniques can be used to determine a protein's structure, this powerful technique has been limited by problems encountered in obtaining high-quality crystals well ordered and large enough to yield precise structural information.

Protein crystals grown on Earth are often small and flawed. The problem associated with growing these crystals is analogous to filling a sports stadium with fans who all have reserved seats. Once the gate opens, people flock to their seats and in the confusion, often sit in someone else's place. On Earth, gravity-driven convection keeps the molecules crowded around the "seats" as they attempt to order themselves. Unfortunately, protein molecules are not as particular as many of the smaller molecules and often are content to take the wrong places in the structure. As would happen if you let the fans into the stands slowly, microgravity allows the scientists to slow the rate at which molecules arrive at their seats. Since the molecules have more time to find their spot, fewer mistakes are made, creating more uniform crystals.

Protein crystal growth experiments were first carried out by the investigating team during STS 51-D in April 1985. These prototype experiments were flown four times and were primarily designed to test vapor diffusion techniques and sample handling apparatus.

The STS-26 PCG was the first controlled or systematic experiment to grow useful crystals by vapor diffusion in microgravity within a thermal control enclosure – the Refrigerator/Incubator Module (R/IM). This equipment was also flown aboard STS-29 and STS-32. Crystals were grown at cold temperatures for the first time on STS-32, demonstrating the potential for using longer flights to process certain proteins.

Results from these experiments have been encouraging, with high quality crystals developing from several of the samples flown. Generally, these crystals are of exceptional size and/or quality when compared to control samples grown in gravity.

During the STS-31 mission, 60 different PCG experiments will be conducted simultaneously using 12 different proteins. These proteins are:

*Isocitrate Lyase – a target enzyme for fungicides. Better understanding of this enzyme should lead to more potent fungicides to treat serious crop diseases such as rice blast.

*Porcine Pancreatic Phospholipase A2 – an enzyme associated with many human disease states including rheumatoid arthritis and septic shock. Successful structure analyses of phospholipase crystals may lead to development of drugs to treat these conditions.

*Human Gamma Interferon (GIF-D) – an enzyme which stimulates the body's immune system and is used clinically in the treatment of cancer.

*Human Serum Transferrin – the major iron transport protein in human serum. It transports iron from storage sites to hemoglobin synthesizing red blood cells and also is a necessary component in media for cell growth.

*Porcine Pancreatic Elastase – an enzyme associated with the degradation of lung tissue in people suffering from emphysema. A better understanding of the enzyme's structure will be useful in studying the causes of this debilitating disease.

*Type IV Collagenase – an enzyme obtained from snake venom (haemmioragic), it is related to collagenase secreted by invasive cancer cells.

*Canavalin – the major storage protein of leguminous plants such as beans and peas, and a major source of dietary protein for humans and domestic animals.

*Malic Enzyme – an enzyme isolated from nematodes. Characterizing the structural differences between it and the mammalian version could lead to the development of an anti-parasite drug.

*Anti-HPR Fab fragment/Fab – the detailed structure would provide a picture of an antibody binding site which recognizes a bacterial "foreign" protein antigen. By learning what antibody binding sites look like, we may better understand how antibodies function in the immune system.

*Factor D – an enzyme necessary for activation of a part of the immune system which plays an important role in host defense against pathogens.

*Turkey/Quail Lysozyme – Sugars are often found associated with proteins, and these sugar/protein interactions are fundamental in all the processes of living organisms. However, very little is known about these interactions.

*Carboxyl Ester Hydrolase – an enzyme which catalyzes the breakdown of carboxylic acid esters like those found in fats. Understanding how this enzyme functions will be valuable in learning how fats and related molecules are made and metabolized.

Shortly after achieving orbit, a crew member will combine each of the protein solutions with other solutions containing a precipitation agent to form small droplets on the ends of double-barreled syringes positioned in small chambers. Water vapor will diffuse from each droplet to a solution absorbed in a porous reservoir that lines each chamber.

The loss of water by this vapor diffusion process will produce conditions in the droplets that cause protein crystals to grow. The samples will be processed at 22 degrees C, as on STS-26 and STS-29.

Just prior to descent, the mission specialist will photograph the droplets in the trays. Then all the droplets and any protein crystals grown will be drawn back into the syringes. The syringes then will be resealed for reentry. Upon landing, the hardware will be turned over to the investigating team for analysis.

The PCG experiments are sponsored by NASA's Office of Commercial Programs and the Microgravity Science and Applications Division with management provided through Marshall Space Flight Center (MSFC), Huntsville, Ala. Richard E. Valentine is mission manager and Blair Herron is PCG experiment manager for Marshall.

Dr. Charles E. Bugg, director of the Center for Macromolecular Crystallography, a NASA-sponsored Center for the Commercial Development of Space located at the University of Alabama-Birmingham (UAB), is lead investigator for the PCG research team.

The STS-31 industry, university and government PCG research investigators include DuPont de Nemours & Co.; U.S. Naval Research Laboratory; BioCryst, Inc.; Schering Plough Corp.; Georgia Institute of Technology; Vertex Pharmaceuticals; Texas A&M University; University of California at Riverside; The Upjohn Co.; National Research Council of Canada; UAB Center for Macromolecular Crystallography; Laboratoire de Cristallographie et Cristallisation de Macromoles Biologiques – Faculte, Nord, Marseille, France; and Eastman Kodak Co.

3.4 SPACE INDUSTRIALIZATION

We all know that the space program requires a substantial financial investment. But if you follow the annual congressional debates about NASA's budget you will note that some of our political representatives seem to regard this as a mere waste of money.

Nothing could be farther from the truth: even today the space program is not only paying for itself, but is producing additional wealth for each and every one of us. Studies have shown that each dollar invested into the space program ten years ago increased today's gross national product (the total value of all produced goods and services) by four dollars. A substantial part of our high-tech industries owe their existence directly or indirectly to the space program. Take, for instance, the rapid growth which our computer industry experienced as a result of the Apollo lunar landing project. Landing a man on the moon as proclaimed by president Kennedy required the development of new miniaturized electronic components, and the first pocket calculators and mini-computers which reached the market by the end of the sixties were a by-product of this research.

The spin-offs of space technology are so numerous that only a few examples can be mentioned here:

Communications satellites, for instance, have developed into a multi-billion dollar industry which turned intercontinental telecasts and phone calls into a normal part of our everyday life while simultaneously leading to a massive reduction in costs. Earth observation satellites have revolutionized meteorology. They provide early hurricane warnings, measure air-pollution and search for mineral resources. Skylab, the first U.S.-space station (total costs 24 billion dollars) discovered previously unknown ore deposits which are valued at 15 billion dollars – just on U.S.-territory alone! Satellite photos enable oil tankers to use ocean currents in order to chart an energy saving course thus lowering their operating costs by millions of dollars. And collisions between ships and icebergs were substantially reduced due to satellite-based navigation and observation systems.

Even developing countries profit directly from space technology: Indonesia, for instance, was able to link its numerous islands via satellite at a fraction of the costs for underwater communication cables. Brazil is using photos made by earth observation satellites in order to develop its unaccessible Amazon territory, which results in considerable savings compared to airborne reconnaissance. In Africa satellites help to track devastating swarmes of insects thus increasing the effectiveness of pest control measures and limiting the amount of insecticides being used.

Agriculture in the U.S. is profiting in numerous ways from our space program: satellites provide early storm warnings, evaluate the soil composition, and check water resources and air pollution. Space-based observation also helps citrus farmers in Florida to protect their plants during cold winter-nights. In order

continued on page 50

Fig. 6 SKYLAB, the first U.S. Space Station, was orbited in 1973. Its three crews spent a total of 513 days in space. They conducted a multitude of scientific experiments, and you can repeat one of their crystal growing experiments with the SPACE AGE CRYSTALS® kit no. 6434.

Fig. 7 The Russian Space Station MIR was placed into orbit in 1986. Staffed year round, this first truly operational Space Station was designed to grow. In 1990, a new 20-ton module named KRISTALL was added to the MIR complex, containing six furnaces for the production of new alloys and super-pure semiconductor crystals.

Fig. 8 Earth-grown crystals of the enzyme isocitrate lyase are badly deformed. Hundreds of experiments performed in earthbound laboratories always produced these dendritic, tree-like crystals.

Fig. 9 Space-grown crystals of isocitrate lyase grown on STS-26. Obtaining uniform crystal prisms like these is essential for studying their structure with X-ray diffraction techniques.

Fig. 10 Space-grown protein crystal of the enzyme porcine elastase. It is 20% larger than comparable crystals grown on earth, and extremely uniform. Understanding the structure of this enzyme could lead to cures for emphysema – a debilitating lung disease.

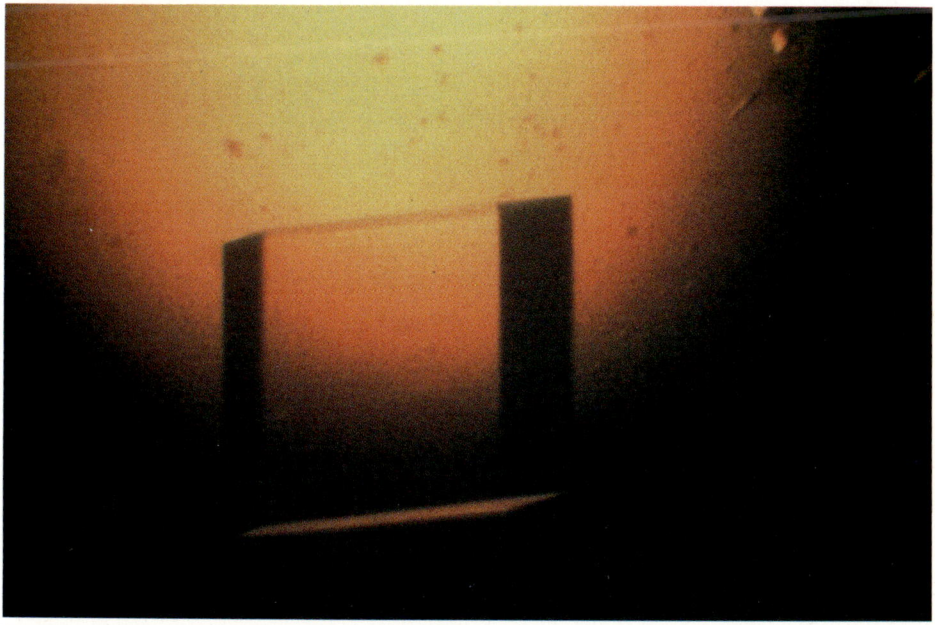

Fig. 11 Space-grown crystal of gamma interferon, an enzyme which stimulates the body's immune system and is used in the treatment of cancer. This crystal is 50% larger than the best crystal obtained in several thousand experiments on earth.

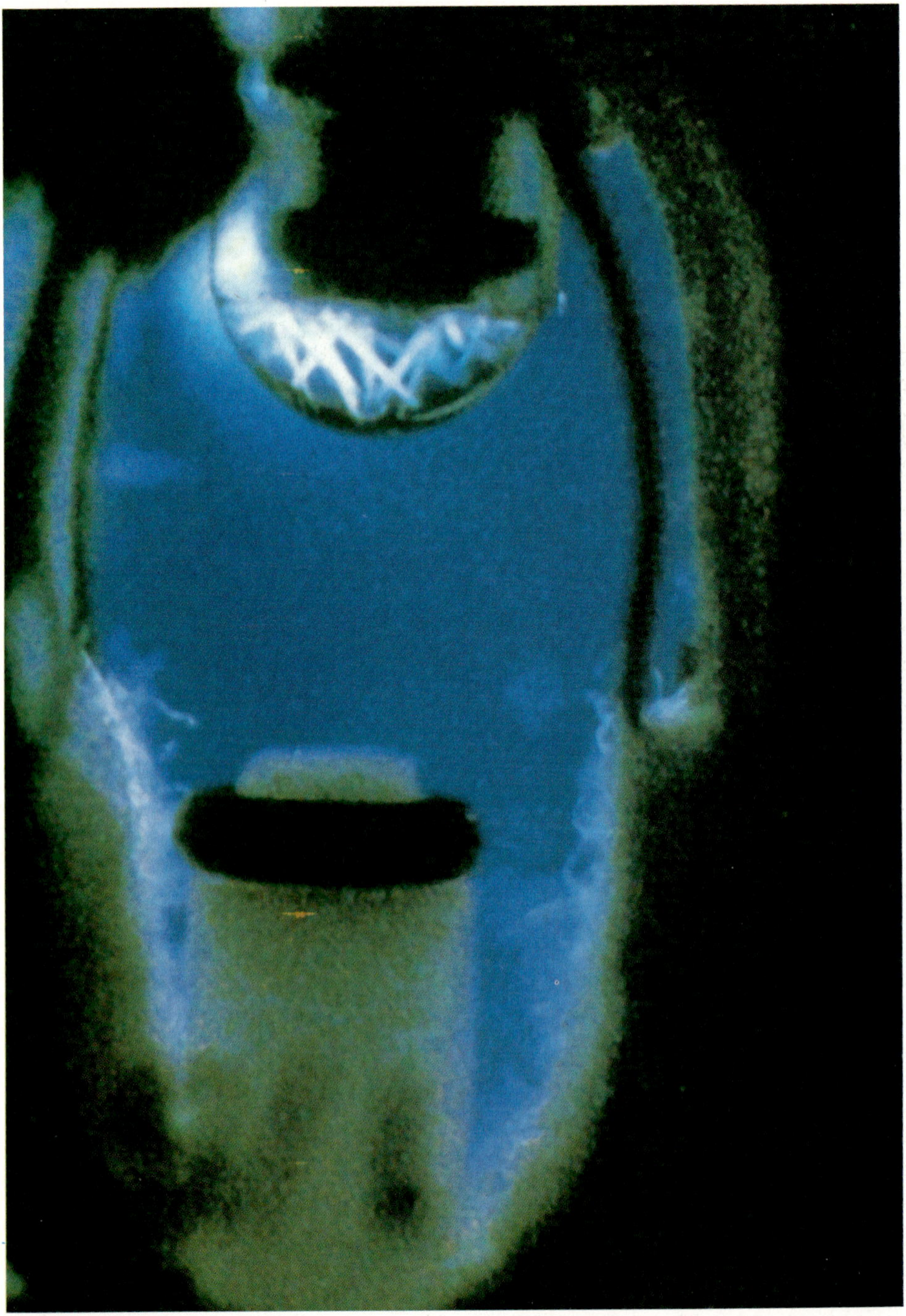

Fig. 12 Closeup of a tray with crystal growing chambers taken by astronaut George Nelson during STS-26 (see also photo on back cover). It shows tiny protein crystals which grew inside a droplet of solution.

Fig. 13 Orbital construction tests which were performed during the Shuttle mission 61-B in December 1985 proved man's ability to erect large structures in a weightless environment.

Fig. 14 "ISLAND ONE" – a space colony of the 21st century housing 10,000 inhabitants. Structures of this size will be built almost entirely of nonterrestrial materials – either from the moon, or from asteroids.

Fig. 15 The interior of this "ISLAND ONE" space colony is designed to provide living conditions similar to those on Earth – with rivers, mountains, lakes and forests – but without Earth's pollution and overpopulation. The colony rotates, thus producing artificial gravity.

Fig. 16 "ISLAND THREE" – a later and much larger space colony with a diameter of 4 miles, and a length of 20 miles. Cylinders of this size can house up to one million people. They will initially be positioned in Earth's vicinity, but later spread into the whole solar system.

Fig. 17 The interior of an "ISLAND THREE" cylinder is much more spacious than an "ISLAND ONE" colony – with wide, sweeping panoramas. Large mirrors reflect sunlight into the interior of the cylinder, and a totally closed environmental life support system recycles all of the colony's waste products.

Fig. 18 Within a few decades, huge SOLAR POWER SATELLITES will collect pollution-free energy from a fusion reactor that has been in operation for about 5 billion years: our sun.

Fig. 19 Astronauts onboard STS 26 grew crystals of lead iodine in an experiment that was part of the National Science Teachers Association's SPACE SHUTTLE STUDENT INVOLVEMENT PROGRAM (see chapter 4.3).

Fig. 20 Forty students (ages 9 to 17) from 11 different schools of the California Central Coast will conduct experiments onboard the Shuttle in 1992. They will study subjects like crystal growth, osmosis, cold fusion, and on-orbit fluid transfer. This photo shows some of the students, teachers, engineers and parents who are involved in this project. A NASA "Get Away Special" canister in the center of the picture houses the experiments.

Fig. 21 Shuttle Student Experimenter Terri Hargis, age 17, during a televised press conference in which she described her crystal growing experiment "BENZOIC ACID CRYSTALS IN SPACE".

Fig. 22 Shuttle Student Experimenter Paul Collier, age 13, who contributes two crystal growing experiments to the California Central Coast "Get Away Special" canister ("KCP CRYSTAL GROWTH" and "ROCK CRYSTALS IN SPACE").

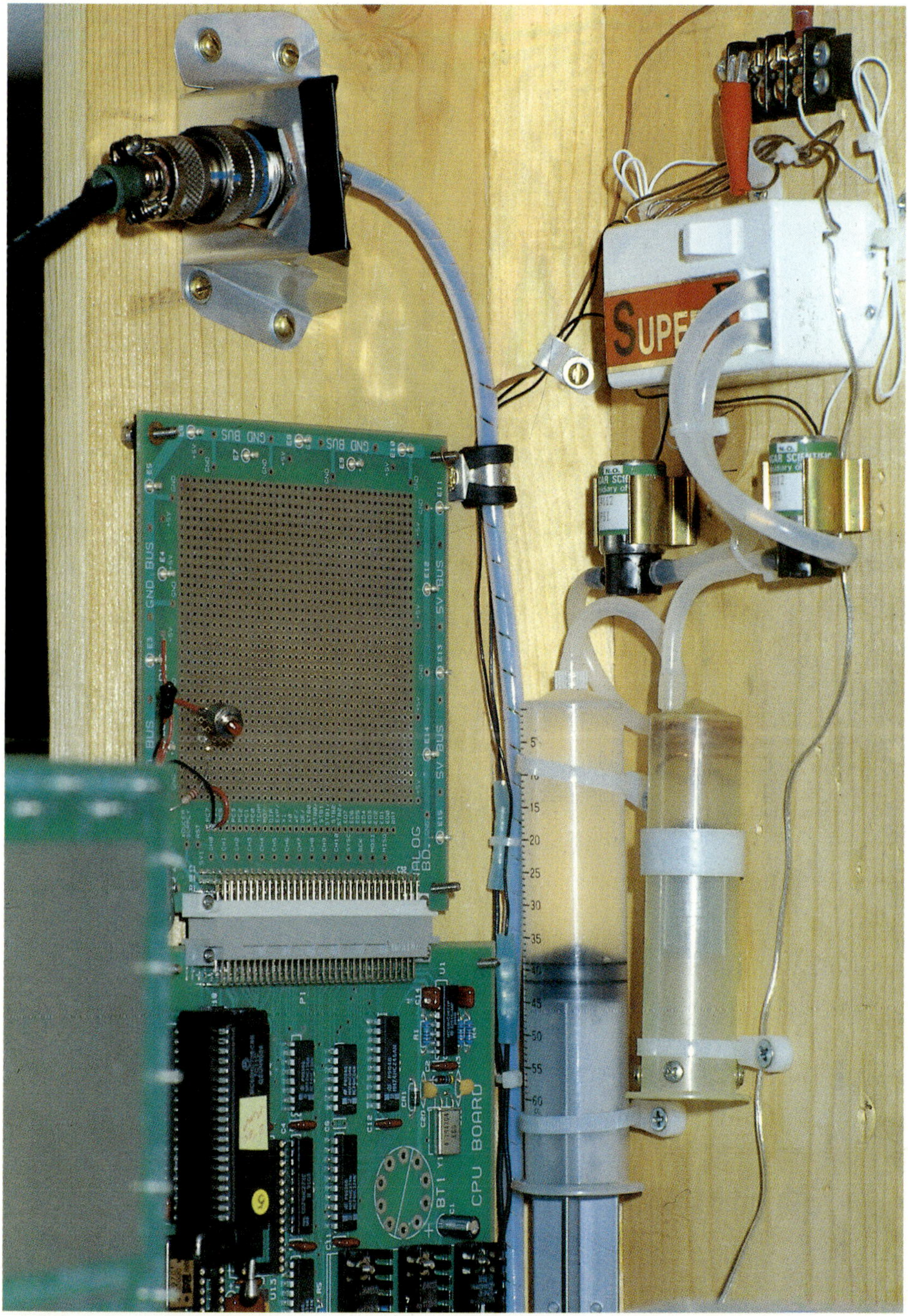

Fig. 23 Experimental set-up for one of the California Central Coast crystal growing experiments that will fly onboard the Shuttle. Using standard off-the-shelf items like plastic syringes or small electric "Hobby Shop" motors helps to reduce the costs for these experiments.

Fig. 24 This laboratory grown quartz crystal has the same chemical composition and physical properties as natural quartz. It forms at very high temperatures and pressures. Slices of such a crystal are inside your quartz-watch.

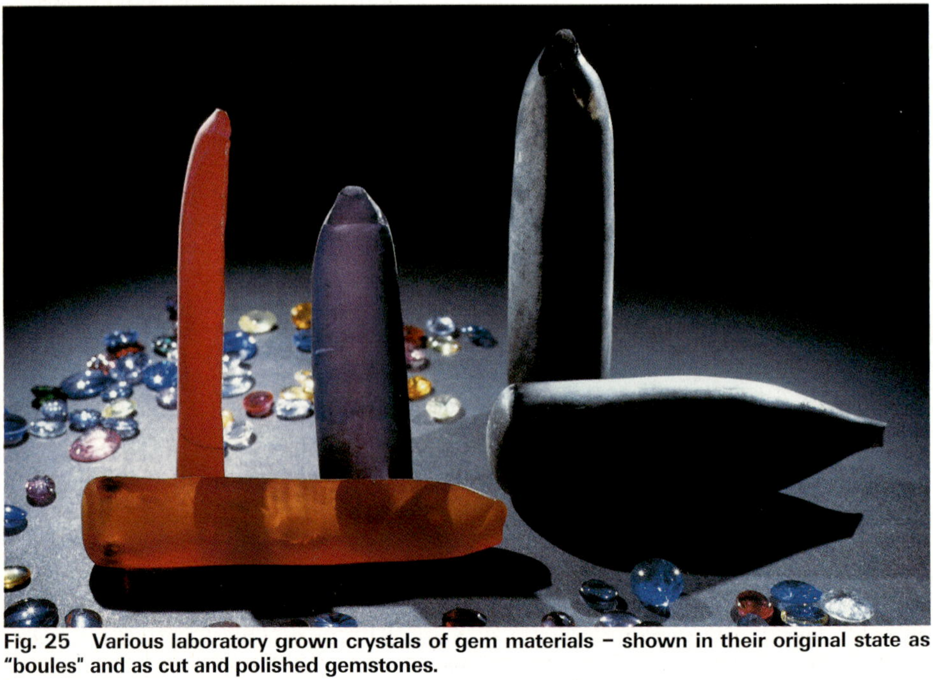

Fig. 25 Various laboratory grown crystals of gem materials – shown in their original state as "boules" and as cut and polished gemstones.

Fig. 26 This metal crystal was grown by melting and recooling bismuth (Bi). Its iridescent colors result from thin layers of bismuth oxide on the crystal's surface.

Fig. 27 Crystals of silicon carbide (carborundum, SiC). These crystals form inside blast-furnaces. They approach the hardness of diamond, and are therefore used as abrasives.

Fig. 28 Many different chemicals produce interesting crystals. Here those of potassium ferricyanide, $K_3Fe(CN)_6$ which were grown from an aqueous solution.

Fig. 29 Solution-grown crystals of potassium ferrocyanide, $K_4Fe(CN)_6$. Note how different these crystals are from those of potassium ferricyanide– although their chemical composition is very similar.

Fig. 30 From the outside, a natural geode looks just like an ordinary boulder. It is, however, unusually light, and when split open ...

Fig. 31 ... it reveals a miniature "crystal cave" which formed when a gas bubble inside molten rock left a cavity in which later minerals crystallized.

SPACE AGE CRYSTALS®

CRYSTAL GROWING KIT

Item no. 657

"AMETHYST GEODE"

Fig. 32 The name of the various types of SPACE AGE CRYSTALS® are those of similar natural mineral specimens. Their chemical composition and physical properties, however, are different from those of their natural counterparts.

Geodes are mineral specimens that – from the outside – look like ordinary boulders which were rounded in a river. They are, however, unusually light since they have a cavity inside, which is often partially filled with crystals.

This kit grows amethyst-like crystals inside two 6-inch hollow geode shells. In addition to the crystal growing substances this kit contains material for producing the geode shells inside a plastic mold.

SPACE AGE CRYSTALS®

CRYSTAL GROWING KIT

Item no. 655

"EMERALD GEODE"

Fig. 33 This kit grows emerald-like crystals inside two 6-inch geode shells.

SPACE AGE CRYSTALS®

CRYSTAL GROWING KIT

Item no. 6533

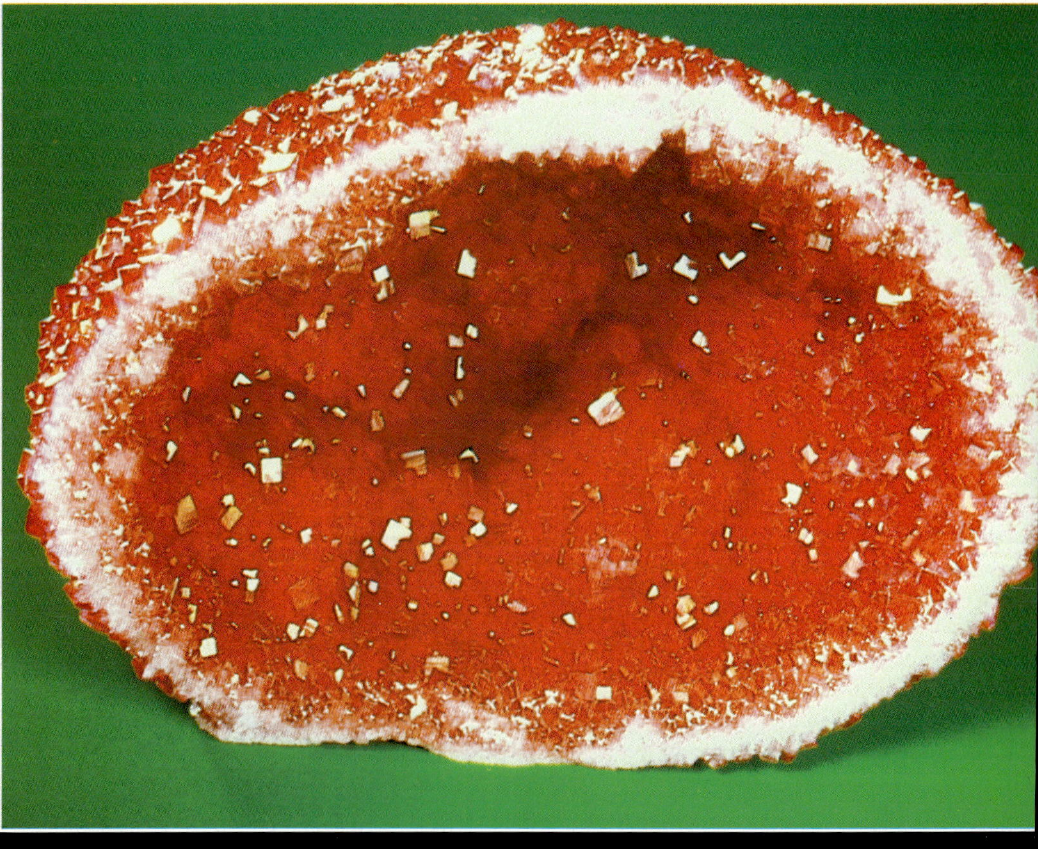

"RUBY GEODE"

Fig. 34 This kit grows ruby-like crystals inside two 6-inch geode shells.

SPACE AGE CRYSTALS®

CRYSTAL GROWING KIT

Item no. 6438

"RED QUARTZ"

Fig. 35 These crystals are grown in a two-step process. First clear quartz-like crystals are grown which then acquire their red tips inside a second solution.

SPACE AGE CRYSTALS®

CRYSTAL GROWING KIT

Item no. 6437

"AMBER"

Fig. 36 This kit grows crystals with the color of amber – a prized gem material which consists of fossilized tree resin, often containing inclusions of insect or plant fossils.

SPACE AGE CRYSTALS®

CRYSTAL GROWING KIT

Item no. 6436

"AQUAMARINE"

Fig. 37 Natural Aquamarine – a valued gemstone – is the blue-green variety of beryl. One completely clear aquamarine crystal found in Brazil weighed a total of 243 pounds.

SPACE AGE CRYSTALS®

CRYSTAL GROWING KIT

Item no. 6435

"ROSE QUARTZ"

Fig. 38 Natural rose quartz is a variety of quartz which rarely occurs well crystallized. Its pink color is probably due to minute amounts of manganese.

SPACE AGE CRYSTALS®

CRYSTAL GROWING KIT

Item no. 6434

"CLEAR COLUMN"

Fig. 39 This kit grows particularly impressive multi-facetted crystals. The SKYLAB astronauts aboard our first space station grew these crystals, and you can repeat their experiments!

SPACE AGE CRYSTALS®

CRYSTAL GROWING KIT

Item no. 6433

"RUBY"

Fig. 40 Natural Ruby consists of aluminium oxide, Al_2O_3. One in 5,000 aluminium atoms is replaced by a chromium atom which causes the red color of this gemstone.

SPACE AGE CRYSTALS®

CRYSTAL GROWING KIT

Item no. 6432

"TOPAZ"

Fig. 41 Natural Topaz is a valued gemstone which is composed of aluminium silicate and fluorine. It forms during the crystallization of igneous rocks.

SPACE AGE CRYSTALS®

CRYSTAL GROWING KIT

Item no. 6431

"CITRINE"

Fig. 42 Natural Citrine is a semiprecious gem. Its chemical composition is identical to quartz, SiO_2, except for a small amount of hydrous iron oxid which causes its yellow or brownish color.

SPACE AGE CRYSTALS®

CRYSTAL GROWING KIT

Item no. 649

"DIAMOND"

Fig. 43 Natural Diamond is composed of pure carbon, and is the hardest natural substance known. Its extreme hardness, brillance and fire make Diamond unsurpassed as a gem.

SPACE AGE CRYSTALS®

CRYSTAL GROWING KIT

Item no. 648

"AMETHYST IN DIAMOND"

Fig. 44 With this kit you can grow "phantom crystals": purple single crystals inside larger clear crystals. In an additional step these crystals can then be grown onto a piece of rock.

SPACE AGE CRYSTALS®

CRYSTAL GROWING KIT

Item no. 647

"AMETHYST"

Fig. 45 Natural Amethyst is a variety of quartz, SiO_2, and is valued as a semi-precious gem. Its color results from small impurities of iron and manganese.

SPACE AGE CRYSTALS®

CRYSTAL GROWING KIT

Item no. 646

"FLUORITE"

Fig. 46 Natural Fluorite consists of calcium fluoride, CaF_2. It occurs in many different colors, and its crystals form cubes and octahedra.

SPACE AGE CRYSTALS®

CRYSTAL GROWING KIT

Item no. 645

"EMERALD"

Fig. 47 Natural emerald is a bright-green variety of beryl [berylium aluminum silicate, $Be_3Al_2(SiO_3)_6$] and a highly valued gemstone.

SPACE AGE CRYSTALS®

CRYSTAL GROWING KIT

Item no. 644

"CALCITE"

Fig. 48 Natural calcite consists of calcium carbonate, $CaCO_3$. It is a widely distributed mineral that forms a great variety of crystals.

SPACE AGE CRYSTALS®

CRYSTAL GROWING KIT

Item no. 643

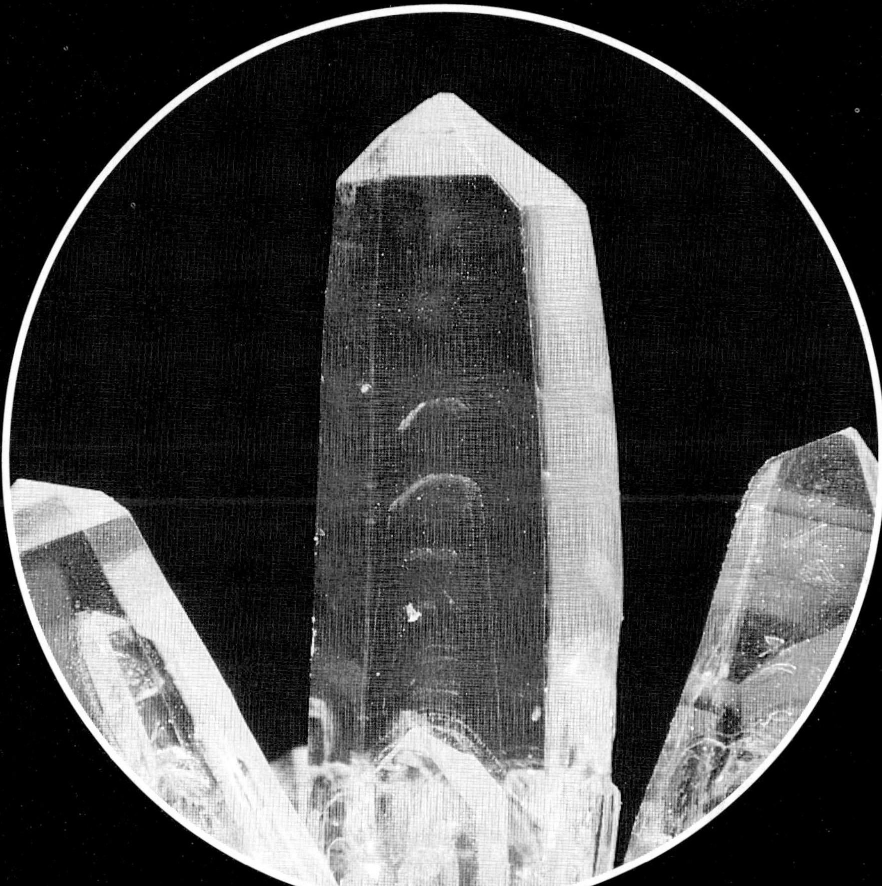

"QUARTZ"

Fig. 49 Actual quartz consists of silicon dioxide, SiO_2. It is part of nearly all of Earth's surface rocks. In its well-crystallized, transparent and colorless form it is also called "rock crystal".

to keep heating costs to a minimum they use temperature charts provided by weather satellites which are updated every 30 minutes. This way an average of 45 million dollars was saved during the last winters. New synthetics, heart pacemakers, improved solar cells, fiber optics which enable physicians to visually inspect inner organs ... the list of spin-offs from our space program could easily be expanded to several thousand examples. But as impressive as this seems to us today – it is just the beginning of the forthcoming era of space industrialization and commercialization. The first permanently manned space stations will open the door for the construction of orbital factories, where new materials will be produced on a regular basis.

But even this is just the prelude to the truly revolutionary changes which will take place in the 21st century. In order to understand their necessity we should remember how small, how limited and fragile earth's biosphere really is: if we would shrink our planet to the size of a globe our atmosphere would be reduced to the thickness of a piece of paper.

The fascinating photos of spaceship earth which the Apollo astronauts brought back from their voyages to the moon made us fully aware of the fragility of our biosphere. And this happened – by chance or necessity? – just at that point in time where an ever increasing pace of industrialization and an explosive population growth started to change our environment in a way that seriously threatens all life on this planet. Particularly grave ecological problems result from a steadily growing demand for energy which of course forms the basis of the wealth of all industrial nations. And a further increase in the world's total energy consumption is inevitable due to our moral and political obligation to enable those countless millions of people who live in poverty and on the brink of starvation around this world to lead a decent human life. To make things even worse the human reproduction rate has reached such stunning proportions that each day 250,000 additional hungry mouths have to be fed, which in the few years remaining until the end of the century will increase the global population by one billion, and most of them will live in third world countries which even today are close to economic collapse.

But how can we possibly produce the required additional energy if even on today's level our environment is seriously endangered? Fossil fuels pollute the atmosphere, and atomic energy has grave technical, political and ecological drawbacks. And even if we were able to solve the problems of controlled nuclear fusion we would still endanger the biosphere due to the inevitable production of waste heat which accompanies every process of energy conversion.

The only solution to this problem are huge solar power plants which will be located in a "geostationary orbit". At a distance of about 36,000 km (22,500 miles) from the earth a satellite requires 24 hours for one orbit. And since the earth rotates just once around its axis during this time a geostationary satellite seems to stand still if viewed from the ground. In this orbit free and unlimited energy is available from a huge fusion reactor which has been in operation for about 5 billion years, and which has still enough fuel left to continue its operation for another 5 billion years: our sun. Solar power satellites (SPS) will collect this free energy 24 hours a day – eventually by using solar cells made out of grown silicon crystals! – and transmit it to earth-bound receiving stations by means of microwaves.

Due to the weightlessness in outer space the construction of huge collector areas with a size of several square kilometers poses no really serious technical problems once the raw materials are in place. Fig. 18 shows an illustration by Boeing engineer and renowned space artist John Olson who depicts the construction of a solar satellite: a space shuttle orbiter docks at the facility's assembly bay. To the left, an upper stage of a heavy lift launch vehicle approaches the facility to discharge its cargo. Note the web-like structures in the foreground which are so lightweight that they would be crushed if used on earth. During the shuttle mission 61-B in December 1985 astronauts already performed construction tests of such elements under weightless conditions (see Fig. 13). Despite the lightweight construction of an SPS it would nevertheless have a total mass of about 100,000 tons which means that if the present shuttle were used to transport the material into orbit a total of 4,000 flights would be required – an unrealistic and far too costly assumption. But fortunately the necessary raw materials do not have to come from earth since there are other, practically unlimited resources in space from which material can be transported into earth orbit in a far more economical way. First of all there is the moon whose composition is known quite well due to the Apollo lunar landings. It consists of 40% oxygen (for breathing and in connection with hydrogen as drinking water and rocket propellant!), 20% silicon (for solar cells), and 20–30% of various metals (for instance as building material for an SPS). True – the distance from a geostationary construction site to earth's surface is just one tenth of the distance to the moon. Movements in space, however, do not encounter frictional energy losses as they do on earth, and therefore it is not distance itself that counts but only the acceleration which is required to get from one point to an-

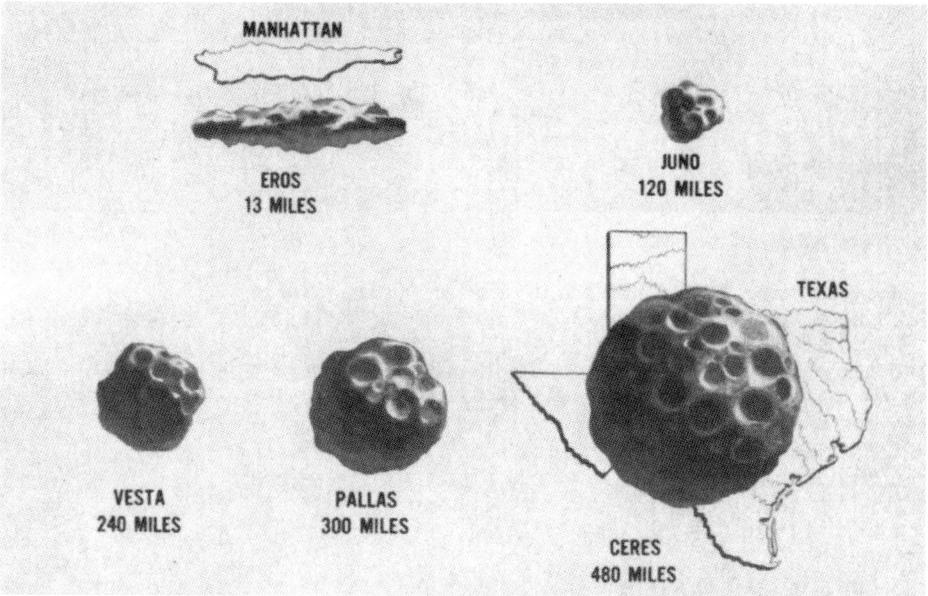

Fig. 50 In March of 1990 an asteroid about the size of an aircraft carrier passed near Earth – less than 400,000 miles away. Earth had been at that point in space a mere 6 hours earlier. Had that asteroid hit our planet the energy released would have been equivalent to 1,000 to 2,500 tons of TNT. An impact in a high population area could have killed millions of people.

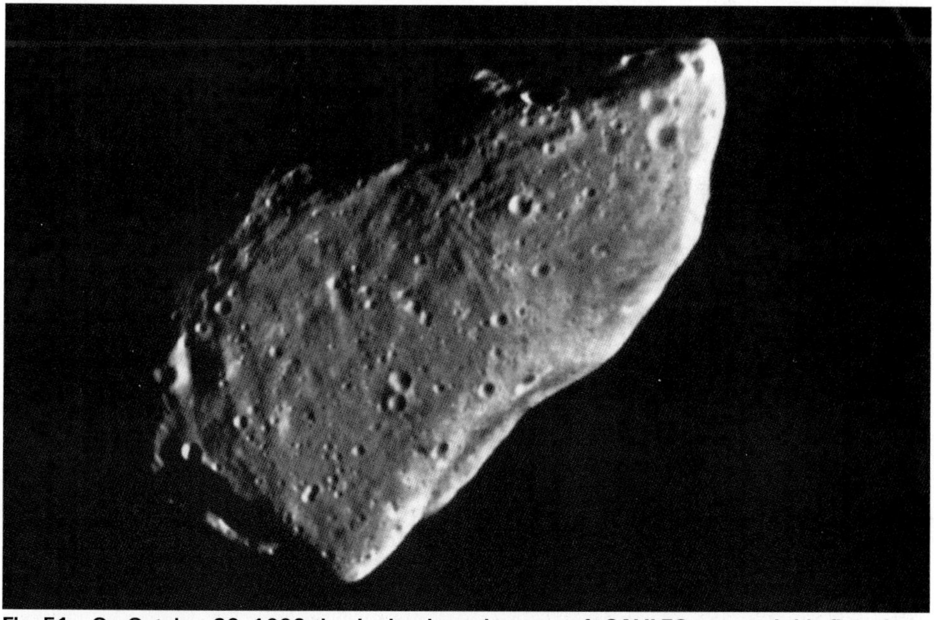

Fig. 51 On October 29, 1990 the Jupiter-bound spacecraft GALILEO snapped this first close-up image of an asteroid (951 GASPRA, size 12 × 7.5 × 7 miles). Many asteroids consist mainly of high-grade nickel-iron steel – worth trillions of dollars. They will supply the raw-material for tomorrow's orbital factories and space colonies.

other. The gravitational force of the moon is much smaller than that of the earth (1/6 g), and there is, of course, no atmosphere which could slow a spacecraft down. Consequently the energy which is required to get from the moon to a geostationary orbit is just 1/20th of the energy consumption for a shipment from earth.

Another practically unlimited source of raw materials are the asteroids, those billions of rocks and metallic objects which circle the sun between the orbits of Mars and Jupiter. About 3,000 of them have a diameter of more than 20 km (12.5 miles), and due to the analysis of meteorites we have fairly good ideas about their composition. A nickel-iron asteroid, for instance, with a diameter of 100 m (330 feet) contains about four million tons of high-grade steel. In all several hundred billion tons of steel should be available from the asteroid belt — and not only from there, since some types of asteroids get so close to earth orbit that the energy requirements for transporting them into geostationary orbit are smaller than those for shipments from the moon! The solar power satellites which could be built out of a single asteroid with a diameter of 300 m (1,000 feet) and a mass of about 50 million tons would have a total energy output that equals today's global production of electricity.

Such huge construction projects require of course adequate accomodations for all those involved. And so, during the next century, space stations will evolve into space colonies. Today's technical literature already discusses plans for the construction of huge cylinder-shaped space habitats with a length of up to 32 km (20 miles) and a diameter of 6 km (3.75 miles). Artificial gravity will be produced by rotation, and the cylinder's inner surface could be landscaped into an earth-like environment with mountains, rivers, lakes and forests. Millions of people will leave our overpopulated home planet to live and work in space. What the Russian teacher and space pioneer Konstantin Tsiolkovsky envisioned eighty years ago will finally become reality for them: "Earth is the cradle of mankind, but man cannot stay in his cradle forever."

3.5 THE REAL SPACE RACE

The National Commission on Space, formed by President Reagan in response to 1984 Congressional legislation, studied existing and proposed U.S. space activities. Highlights of their recommendations were mining operations of the moon and asteroids, interplanetary factories and spaceports as well as human colonies on the moon by 2017 and on Mars by 2027. Their report — entitled PIONEERING THE SPACE FRONTIER — OUR NEXT 50 YEARS IN SPACE — is available in book form.

The following is excerpted from testimony presented by Ben Bova, former president of the National Space Society, to the National Commission on Space, at their hearings in Boston on October 28, 1985 (Source: National Space Society, SPACE WORLD magazine, December 1985):

"Today I would like to speak about the fundamental reason why the space program is crucially vital to the American people, and to the people of the entire world. It is important to understand the basic imperative that is driving the human race's expansion into space.

To survive on Earth, we must expand into space.

The matter is just that simple, and just that important. We live on a world with a population of more than four and a half billion human beings. A quarter of a million babies are born every day: most of them in the regions of the Earth that are poorest, hungriest, and least educated. Every day we see fresh evidence of the enormous tensions that rack our world: wars and threats of wars, terrorism, famine, revolution.

As world population continues to grow, world problems intensify. Imagine the tensions and terrors of a world of six billion human beings, which is what the global population will be in little more than ten years. Eight billion. Ten billion.

Our growing world population is using up the Earth's resources at a fearsome rate. Primeval tropical forests are being devoured; vast regions of the oceans that once were rich fishing areas are now becoming barren; resources of fuels and minerals are being depleted.

As long as population growth continues, the Earth's supply of resources will continue to be gutted. And long before we reach the limits of physical supply, tensions over the price and distribution of these resources will make the threat of starvation, terrorism and war even more acute.

As long as the human race is confined to this one planet, our growing population will threaten us with disaster. There are three possible solutions to this fundamental problem.

One: Do nothing. Allow the Four Horsemen of the Apocalypse to perform their historic roles. In a

world armed with roughly twenty thousand megatons of nuclear weaponry, this could mean the end of civilization, perhaps the end of all human life.
Two: Enforce population control measures all across the world. Quite frankly, this would entail a dictatorship so crushing that, even if it could be accomplished, it would mean the end of human freedom.
Three: Stop restricting ourselves to the resources of this one planet. If Earth cannot supply the energy and natural resources that we need, then we must find those resources elsewhere.
Recognize that throughout history, population explosions have been stopped either by natural disasters or by raising the standard of living. Rich people do not make population explosions: poor people do. Our task is to make people rich. Not by taking wealth from one group to distribute to another. Not by dividing the Earth's finite supply of resources into thinner and thinner slices as the population grows. But by reaching out into space, to find and develop new resources to make the human race's 'pie' larger.
We know the resources are there. The past quarter of a century of space exploration has given ample evidence of that. We are developing the technology to use those resources, and to make the Earth richer. That is the imperative of space, the real reason why we must have a vigorous program of space exploration and development.
We are in a space race today. Not a competition between nations jockeying for prestige and power. This race involves the entire human species in a contest against time. All the people of the Earth are in a desperate race against global disaster.
To save the Earth we must look beyond it, to space. To prevent the collapse of civilization and the end of all human freedoms that America has championed for centuries, we must understand that our planet does not exist in isolation. Our Earth is part of a solar system that is incredibly rich in energy, in natural resources, in all the wealth and raw materials that we need to build a flourishing, fair, and free global society.
This new space race is, in reality, a crucial struggle against humankind's ancient and remorseless enemies: hunger, poverty, ignorance, and war.
We must win this race, for one brutally simple reason: survival."

4.1 COOKING WITH GAS

In an attempt to involve the general public in its mission of space exploration and utilization NASA enables non-scientists – including students – to conduct experiments onboard the Space Shuttle.
Interested parties – like schools, universities, and even individuals – can rent so-called GET AWAY SPECIAL (GAS) canisters, place their experiments inside, and have them flown onboard the Shuttle for a fraction of the actual costs.
In cooperation with the Vandenberg Air Force Base Chapter of the American Institute of Aeronautics & Astronautics THE KRISTAL CORPORATION has provided funding for the GAS # 450 student crystal growing experiments. The following article by Dr. Brenda Forman about this GAS project is reprinted from SPACEWATCH – a monthly publication of the UNITED STATES SPACE FOUNDATION. Dr. Forman is Corporate Director of the Lockheed Corporation, and has been teaching Space Policy at the International Space University.
Education (to paraphrase Mark Twain) is like the weather: everybody talks about it but nobody does anything about it. Our kids score lower in science and math than their peers in any other industrialized country in the world, but so far we've done nothing about the problem but argue – and resist any effort at real reform. So when you find a glowing exception to this sorry situation, it deserves notice. Moreover, it deserves emulation. Most of all, it deserves support.
I encountered just such a glowing example during the Space Week celebrations this July at Vandenberg Air Force Base. Led by the VAFB Chapter of the AIAA (American Institute of Aeronautics and Astronautics), it involves 30 students from 11 schools throughout the California Central Coast, dozens of local industry and government engineers – and one GAS can. The GAS can in question has nothing to do with Texaco; it stands for "Get Away Special" and it's a cylinder you stuff with experiments and which NASA then tucks into one of the spare crannies left unoccupied by the main payload in the shuttle bay. The shuttle can carry about twelve of these on a typical flight.

The whole effort is led by two energetic engineers from the Western Space and Missile Center, Ed Burke and Marty Waldman. This is dedication in action: two talented guys putting their own skills to work to help enthusiastic young kids develop and then actually fly their own experiments in space. "I've seen a lot about education", says Ed Burke. "You need hands-on experience. That's missing in the schools. They're not teaching kids how to do anything!" The Get Away Special Team and its dozens of engineers and science advisors is out to change that.

Burke and Waldman wrote a script for a video tape describing the GAS can project and persuaded the local Air Force Audio-Visual people to produce the tape for them. They sent the tape around to all the schools on the California Central Coast to get students interested in participating in the project. The response was overwhelming: 500 proposals for experiments. In June 1990, they picked the winners: 13 experiments, two and a half times more than anybody has ever before tried to fly in one GAS can. It's a great mixture: from cold fusion to crystal growth in zero gravity, from on-orbit fluid transfer to seeing if glue will bond in space, from a sophisticated osmosis experiment to an inquiry into that truly burning question, how to improve toilets in space.

Students meet once each month at Allan Hancock Community College with the participating engineers. Waldman and Burke tried to choose primarily 9th and 10th graders who will still be in school when the GAS can flies in 1992. (The kids are only worried that they won't get their experiments back from NASA fast enough to make their findings before they graduate.) Engineers also work with kids in their homes, helping to figure out what will work in space, designing flight hardware, finding parts. After a year of intense work, the canister flight structure is done, complete with microcontroller boards and software, and power system.

Some 50 companies around the country have helped out with money and donations but this is at bottom a labor of love for all concerned. (Burked designed the power system, doing a lot of wiring in his own garage.) Some solid science is very likely to come out of it all, but the real purpose is to excite young people about space and science. They're doing just one GAS can now, but Marty Waldman wants to see this become an integral part of the school system across the country. "I'd like to see the GAS can project become as much a part of the school curriculum as football and baseball", he says. "Our structure is completely re-useable. We have backup experimenters still working and thinking. We'd like to be able just to plug in new experiments and refly the canister on future shuttle flights."

Coincidence, fate or just plain good luck is helping out. Several months back, Ed Burke bought a crystal growing kit for his kids and noticed it was made by The Kristal Corp. in Sparks, Nevada. On impulse, he called Kristal and asked them if they were interested in a project where kids were going to grow crystals in space. They were. The upshot was that their President, Heinz Teige, came to the VAFB Space Week, bringing his video crew with him to make a video about the GAS can project for distribution to school systems throughout the world. With any luck, that could start a bandwagon.

4.2 STUDENT EXPERIMENTS FLY ON THE SPACE SHUTTLE

The following article by Edmund D. Burke, GAS # 450 Technical Director, Vandenberg Air Force Base, covers various technical and organizational details of this project. It outlines the challenges that all participating students, teachers, parents, and engineers were confronted with, and thus reflects their remarkable enthusiasm, energy, and dedication.

The GAS #450 engineers and THE KRISTAL CORPORATION are planning future GET AWAY SPECIAL flights, and if you are interested in participating in these projects you can obtain more information by sending a self-addressed stamped envelope to: THE KRISTAL CORPORATION, P.O. Box 1677, Sparks, Nevada 89432, (702)359-0411

INTRODUCTION AND HISTORY

How would you like to have your experiment chosen to fly on the Space Shuttle? Does this sound impossible? Are you thinking that you could never come up with an experiment idea good enough to be flown aboard the Space Shuttle? Terri Hargis, California Central Coast Student Experimenter said, "I never thought my experiment would be chosen, and neither did many other students, but our experiments were selected." Your experiment idea could be selected to fly too if you become involved in a future National Aeronautics and Space Administration (NASA) "Get Away Special" Project sponsored by The Kristal Corporation.

What is the "Get Away Special" program and why was it started by NASA? How do you purchase a canister to fly experiments on the Space Shuttle? The "Get Away Special" (GAS) program was founded by NASA to increase community involvement in the civilian space program, to encourage the younger generation to excel in science and engineering, and to encourage the use of space by all researchers: private individuals and organizations. The standard GAS container provides the means of installing small experiments in the Space Shuttle cargo bay.

An individual or organization can purchase a 5 cubic ft. canister or a 2 1/2 cubic ft. canister, and must design/build an experiment rack that holds the experiments and develop their own electrical power, heating, data handling facilities, etc., within his own GAS container. The experiment rack that you will develop will be inserted into a NASA Cylindrical Pressure Enclosure and will bolt into the insulating cover (as required) that is part of NASA's Cylindrical Pressure Enclosure. The Cylindrical Pressure Enclosure can be maintained at about 1 atmospheric pressure with dry air or dry nitrogen during flight or placed in a vacuum if desired. The experiment rack insert and all experiment hardware must weigh under 200 lbs. for a 5 cubic ft. container and 100 lbs. for a 2 1/2 cubic ft. container. For more information regarding the Get Away Special Program, contact:

Get Away Special Technical Liaison
Code 740
Goddard Space Flight Center
Greenbelt, Maryland 20771
Phone: (301)-286-6760

Ask for a "Get Away Special" (GAS) small self-contained payloads experimenter handbook. This handbook will help you to develop a GAS payload.

NASA began flying GAS payloads during its fourth shuttle flight in June of 1982. Since then, 54 such payloads have been carried into orbit during 14 shuttle flights. The Vandenberg Air Force Base (VAFB) chapter of the American Institute of Aeronautics and Astronautics (AIAA) purchased a "Get Away Special" canister from NASA in 1987 and are allowing California Central Coast

Fig. 52 The Get Away Special Bridge, capable of holding a maximum of 12 canisters, undergoes preparations for installation in the payload bay of the Space Shuttle Columbia.

GET AWAY SPECIAL
SMALL SELF-CONTAINED PAYLOADS
CONTAINER CONCEPT

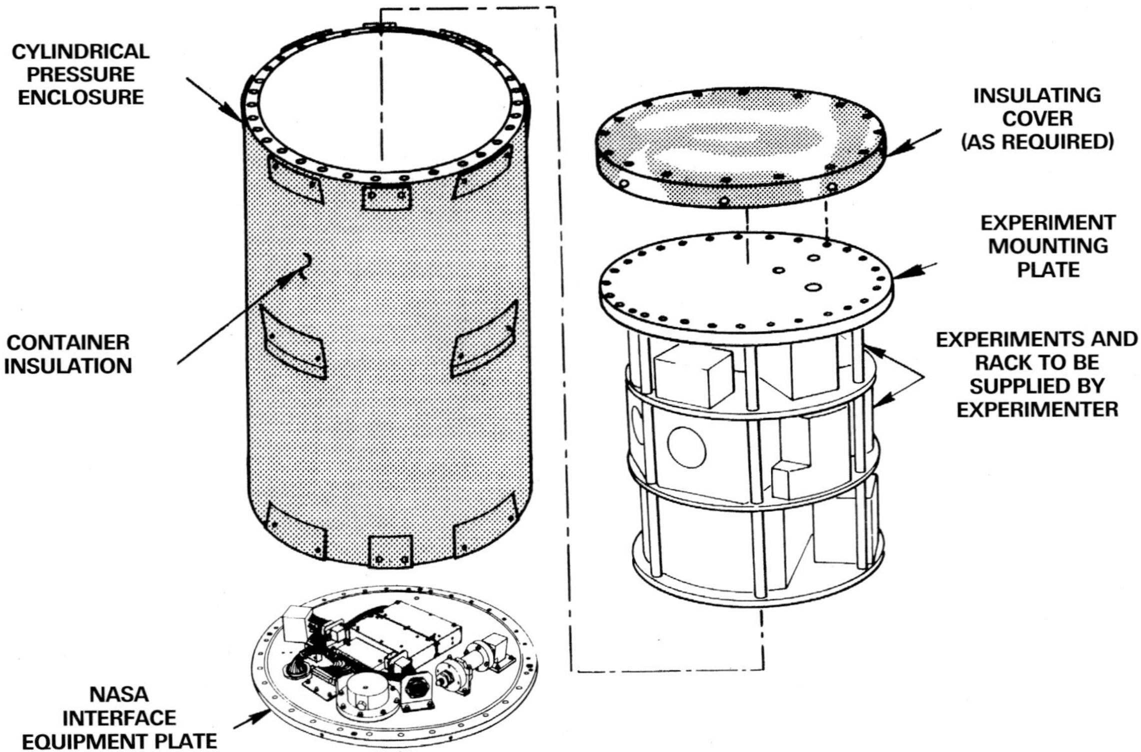

Fig. 53

students an opportunity to fly experiments on the Space Shuttle. The reservation was fully paid to NASA by the VAFB AIAA by raising money in the local community. Our NASA assigned payload reservation number is GAS # 450. The main goal of the project was to increase knowledge and enthusiasm for science in the younger generation in all schools from San Luis Obispo to Santa Barbara, California. Several engineers and scientists who work for ITT/Federal Services Corporation, Lockheed Missiles and Space Company, Tracor Aviation, Computer Science Corporation, Lompoc Aviation, Allan Hancock Community College, the U.S. Government at VAFB and the local community are currently designing the canister. The design team is providing the electrical, mechanical, and thermal expertise needed to complete the system level canister design and provide assistance to the students on their experiments.

At the project outset, a video tape and form was used to solicit the students of the California Central Coast Schools for ideas/proposals. With a response of over 500 applications, the design team ended up choosing the best experiment ideas/proposals on the 10 June 1990. Thirty experiment finalists gave fifteen minute presentations on their experiment ideas. Out of these, fourteen student experiments will fly on the Space Shuttle in 1992. Forty students (ages 9 to 17) from 11 different schools are involved in these 14 experiments.

GAS # 450 PEOPLE CONSIDERATIONS

The California Central Coast "GAS # 450" payload has something for everyone in the local community. The heart of the project consists of the student experimenters. They are all young, talented, are very excited about working on their experiments, and eager for directions in which to channel their energy. The GAS opportunity, in our case, has provided hundreds of people of all sorts, an avenue to apply their creative energy into space research, supposedly available only to the degreed and opportune. Directly supporting the forty students are the teachers and advisors at the schools they attend and the technical staff of engineers/scientists from the local community.

Even with this kind of support, the project could not be successful unless other people pitched in and helped. The student experimenters are young people with very active lives. They have heavy workloads in school. They play athletics, and some even have part-time jobs. The students have but so much time, and to take on a commitment as extensive as the GAS # 450 project means that previously allotted time for certain activities has to be shortened or eliminated. The parents are supporting the experimenters by providing transportation, helping on experiment designs at home, attending meetings, and altering their own schedule during conflicts to insure the experimenters enough time to work on experiments. Friends of the lead experimenters also have an important supporting role. A friend can join in on the enthusiasm of the project by sometimes participating, sometimes just listening, and sometimes by just being there when needed. The sponsors help provide the equipment it takes to complete the experiment. They are the unsung heroes because many of the experiments use specialty items that are not available at local stores and are not available as "off the shelf" items anywhere.

NASA has been supportive from both a personal and organizational standpoint. The technical staff of the GAS #450 project must coordinate every design change through NASA to insure a dangerous experiment is not inadvertently packaged in the GAS canister. The NASA personnel have responded to the AIAA's efforts in a timely manner and a good working relationship has been established between the two organizations. Last but not least, is the community at large. Community media is important as it tells the local community what positive things a few of its citizens are doing and accomplishing. The community will soon learn of the sacrifice that these experimenters are making and give them the "pat on the back" they deserve as well as provide the additional support that is needed through fundraisers and donations. NASA Goddard, parents, family, AIAA, contractors (sponsors), schools, friends and teachers are veins feeding the heart of the project, "the young students". As long as everybody keeps the heart pumping, excited and willing to work hard, the project stays fun and on schedule.

STUDENT SCHEDULES AND MEETINGS

One of the guided objectives is to provide the students with an experience similar to the aerospace industry at large. Meeting organization, action item discipline, scientific procedures, and schedule milestones processes are utilized to enhance the aura about the project. The GAS # 450 used a 3 phase development approach. Under Phase 1, a preliminary experiment design paper, drawing and prototype model is due. Phase 2 development consists of producing a final experiment paper and drawing and a functional prototype model. Professional laboratory notebooks donated by Scientific Notebook, Co. with a California Central Coast Space Shuttle Experimenter engraving is also part of Phase II development. The laboratory notebook is a written record of the student experiments. It is broken into categories which use the Scientific Method, such as: Purpose, Procedure, Experiment Apparatus, Observations and Data, Calculations and Conclusions. The experiment notebook is a vital tool in explaining to NASA in a final report how microgravity and space environmental factors effect the experiment. Plenty of ground testing is needed to prove the purpose of the experiment and draw a final conclusion. Phase 3 development consists of producing the final flight ready experiment module and interfacing it to the system level canister design. The students meet with the GAS # 450 Technical Staff once a month at Allan Hancock Community College in Santa Maria, California to provide a status of their progress and discuss any difficulties in experiment design/development. It has become apparent that regularly scheduled design reviews are crucial for any GASCAN project program. We have found that these reviews identify problems in designs, can help find flaws in basic system approaches to experiments, allow one to make changes to systems before they become difficult to change or a safety problem, help identify safety related issues and provide a good educational experience for the students. The technical staff monitors all three phases of the project each month with a progress chart. Three press conferences with local media and newspapers are scheduled, one at the outset of the project, one after Phase II development is complete (27 June 1991) and one before flight on the Space Shuttle in 1992.

The chassis and electrical design was developed so that it can be reused on future flights. The AIAA group sponsoring the canister intends to continue flying payloads from the California Central Coast and abroad after GAS # 450 is launched in 1992. The chassis was designed in a modular fashion to accomodate a multitude of experiments and to keep technical and fabrication hurdles to a minimum for follow-on flights. The flight ready module for GAS # 450 has been built.
The GAS # 450 electrical, mechanical and thermal design is unique. No other GAS canister has ever flown more than six experiments in one canister. The GAS # 450 canister will fly fourteen experiments. The GAS # 450 design is something our technical staff is very proud of!

THE STUDENT EXPERIMENTS

Payload GAS # 450 is a multidisciplinary package composed of six self-contained experiment modules. Fourteen experiments have been developed by the California Central Coast School students, among them several crystal growth experiments:

Terri Hargis, Cabrillo H.S.
"Benzoic Acid Crystals in Space"
The objective of the experiment is to study the effect of zero gravity on the arrangement of molecules in Benzoic Acid, an organic substance that has a melting point of 70 degrees Celsius.
The Benzoic Acid crystal will be grown on the ground and tested for crystal lattice structure using state of the art cystallographic instruments. The same crystals grown on the ground will be melted in space and allowed to reform under the conditions of zero gravity. Because convection and buoyancy effects are essentially eliminated in space, it should be possible to produce more nearly perfect crystals. The resulting "Space Crystal" will be compared to earth grown crystals after recovery from the Space Shuttle.
The entire experiment will be encased in an old space shuttle thermal tile. Two thermofoil heaters will be used, 46 x 76 mm each, to heat the well slides where the Benzoic Acid crystals are placed. The crystals are placed into the depressions on the well slides and are covered with sapphire slip covers. In space, the heaters will be turned on to allow the crystals to melt, and turned off via the microcontroller once the 70 degree Celsius temperature is reached. A thermal sensor will output an analog voltage to be recorded and used by the microcontroller to deactivate the strip heaters at the right time. A thermal sensor will be in contact with the aluminium tube running through the thermal tile.

Module 6 – Paul Collier, Orcutt Junior Middle School
"KCP Crystal Growth"

The purpose of this experiment is to determine the effects of micro-gravity on the growth of linear conducting crystals. The crystals will be grown by an electrochemical process. Crystals grown under one-g are slightly deformed due to their weight. In space, this complication is eliminated and nearly perfect crystals can be grown. The experiment is an extension of research done on GAS-007 flown on Columbia Mission, STS-61C. The crystal froze once in orbit and the experiment failed.

The crystals to be used are called linear chain conductors. Those to be studied are "salts" formed from a material called tetracyanoplatinate which is abbreviated as TCP. The particular compound which will be studied is partially oxidized potassium tetracyanoplatinate (KCP). This compound exhibits the columinar stacking typical of TCP complexes. When electrical current is passed through the solution, crystals of KCP are formed on the anode (positive terminal). This electrochemical process forms crystals which have 100,000 times better conductivity than crystals formed by other methods. The crystals formed in space will be "straighter" and their conductivity could be much greater. Maybe up to 10 times better; which would produce a crystal with 1,000,000 times the conventional type conductivity. This experiment could ultimately lead to a better understanding of superconductivity.

The crystals will be grown in 2 small plastic cylinders with the crystal solution premixed inside. The cylinders will contain a thin sheet of platinum metal as the cathode and a small pin shaped platinum anode upon which the crystals will grow. A 1.5 VDC voltage will be placed across each cell and the crystal will grow. A temperature sensor will be mounted on the cells. Temperature vs. time data will be stored by the microcontroller. The cells will be well insulated so that they will not freeze. Pictures will not be taken of the cells. The main objective is to return the crystals safely to earth so they can be analyzed.

Eric Edelbrock, May Grisham Elementary
"Silver Crystal in Space"

The purpose of this experiment is to grow silver crystals in microgravity using an electrolytic procedure. Normally, gravity dominates the behavior of fluids, with undesirable effects on quality of crystals. These defects impair the performance of the crystals made for sensors or electronic components. It should be possible to produce more nearly perfect crystals in space. Structural defects in crystals produced in space should be minimal. The silver crystals grown without convection should be less compact and cohesive (more powdered) than those produced on Earth.

The experiment consists of two pressurized growth chambers of silver nitrate solution. When the two-way valve is opened by the microcontroller, the fluid fills the two crystal growth chambers. The excess air is pushed out through the air/orifice filter when fluid enters the crystal growth chamber. The overflow chamber is used just in case some of the silver nitrate solution makes it through the filter. Five volts DV will be placed across the anode and cathode. The anode and cathode electrodes are made out of insulated copper wire. The silver crystals will accumulate on the copper wire between 24 to 48 hours after placing DC voltage across the cell. The temperature is kept above freezing for this time period. A temperature versus time profile is stored via the microcontroller over the duration of the experiment.

Other experiments that will fly on the Space Shuttle in the GAS # 450 canister are as follows: "On-Orbit Fluid Transfer Experiment", Lockheed Missiles & Space Company with local students in Sunnyvale, California; "Effect of Radiation on Bacteria and its Resistance to Househould Chemicals, Lompoc H.S.; "Germination in Space", Lompoc H.S.; "Bacteria in Space", Fesler Elementary School; "Zero Point Energy", Lompoc H.S.; "Bubbles in Space" and "Space Adhesives", San Luis Obispo High School; "Osmosis" and "Reverse Osmosis", Pine Grove Elementary; "Seeds Exposure to Radiation", Paulding Middle School; "Fluid Separation and Surface Tension", Lompoc H.S.; "Earth's Magnetic Field Effect on a Magnetic Sensor", Judkins Intermediate School. All the students and experiments are awesome! The students have become true scientists! The knowledge they have gained by being part of this project will make their lives after the project more rewarding.

4.3 SSIP CRYSTAL GROWTH EXPERIMENT

In the past decade several student crystal growing experiments flew onboard the Shuttle as part of the National Science Teachers Association's SPACE SHUTTLE STUDENT INVOLVEMENT PROGRAM (SSIP). The following NASA paper [CRYSTAL GROWTH USING A SEMIPERMEABLE MEMBRANE (SSIP EXPERIMENT SE82-5)] describes one of these experiments which was originally launched on the Shuttle mission STS 51-L, but was lost when the Challenger exploded. It was launched again on STS 26 in September of 1988.

INTRODUCTION

In 1986, one week after the loss of the Shuttle Challenger, its crew and his experiment, Richard Cavoli sat in the Visitor's Gallery of the United States Congress and heard President Ronald Reagan say: "We see the (American) dream coming true in the spirit of discovery of Richard Cavoli. All his life he's been enthralled by the mysteries of medicine. And, Richard, we know that the experiment that you began in high school was launched and lost last week. Yet your dream lives ... (It is) work of noble note ... work that could reduce the harmful effects of X-rays on patients and enable astronomers to view the golden gateways of the farthest stars."

Richard was in a White House office waiting to see the President when he witnessed the disaster on television. He did not know until later that he was there because President Reagan was to talk about him and his SSIP experiment in the State of the Union message. He had left Florida before the launch to be with his ailing grandmother when the White House called.

Richard himself called the development and integration of his crystal growth experiment "a trail blazing adventure". He commented further, "The most rewarding thing was that there has been no planned path. I conceived the experiment myself and I had to carry it through."

He was a junior a Marlboro Central High School in Marlboro, New York, when his proposal was accepted by SSIP. He excelled further by becoming valedictorian of his high school graduating class. Richard says, "After I get my education, I think I'd consider going into the space program in one respect or another, and it may very well be flying on a Shuttle mission."

Richard credits his interest in science to his early educational experience. "It may have stemmed from the fact that the parochial school I attended had a separate science teacher and by fourth grade we were writing up experiments and doing them ourselves."

Union College in Schenectady, New York sponsored Richard's research. Dr. Charles Scaife, Union chemistry professor, worked closely with Richard as his advisor. Union, Richard's alma mater, paid for transportation and equipment. The college also made available the school machine shop and provided the services of two skilled machinists. "If it weren't for Union", said Richard, "none of this would have been possible." Personal attention, access to faculty, an interdisciplinary approach, and a broad, liberal perspective on the scientific process have been important to Richard.

Richard used the time between the first and second launches of his experiment to refine his hardware and further test his procedures.

Richard's experiment was launched for the second time on STS-26 Discovery, September 1988.

BACKGROUND AND OBJECTIVES

The objectives of the crystal growth experiment are multi-faceted in their applications to science and industry. Foremost is the determination of conditions vital to growth of high purity materials in space. From this platform successful preparation of large, pure, single crystals of lead iodide in the micro-gravity environment could lead to the following benefits:

* Development of better cathodo-luminiscent materials used in X-ray film for dental and medical diagnostics;
* Refinement of image-intensifying screens used in astronomy for the detection of gamma and X-rays;
* Preparation of other inorganic materials in space such as molecular sieves which are used as drying agents in chemical processes, such as purifying a blood sample.

The basic purpose of the experiment, however, was to grow larger, purer lead iodide crystals in the space environment than could be grown on Earth. Cavoli designed the experiment based on the following assumptions:

1. Since lead ions do not diffuse, placing the lead in a specific orientation to the membrane allows for growth to occur on only one side of the membrane.
2. Variations in ionic concentrations affect rate of crystal formation.
3. Supposing that ionic attractions are the only forces acting on the crystals grown through the semi-permeable membranes, morphology of the crystal can be directed by shape of the membrane and pore size.
4. Eddy currents produced by the Earth's gravity, which are believed to cause perturbations in crystal growth, will be absent in the microgravity environment.

Given these parameters, Cavoli hypothesized that crystal growth occurring in the microgravity environment would enable the growth of large, perfect, rod-shaped crystals. Furthermore, the absence of gravitational perturbation would allow for formation of a purer, more insulative crystal.

MATERIALS AND METHODS

Varying concentrations of lead and iodide ions were allowed to diffuse from opposite sides of a cellulose membrane in a microgravity environment. Growth of the resulting lead iodide crystals ocurred on the semi- permeable membrane, with the objective of achieving plate-like, large, rod-shaped crystals.

Five four-chambered, acrylic containers were designed and machined at Union College. According to Charles Scaife, the four chambers were filled through the fill plugs with 0.0624 M lead (II) ace-

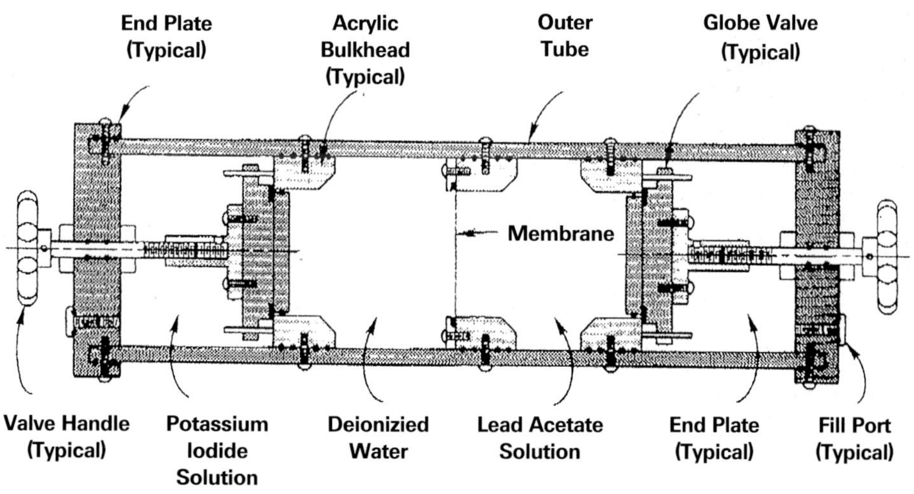

Fig. 54 Cross-section diagram of the crystal growth apparatus.

tate; separated by a valve and the membrane from deionized water. These were further separated by a valve from either 0.360 M, 0.180 M or 0.090 M potassium iodide (KI). The three different KI concentrations were used in the Shuttle experiments with similar control experiments performed on Earth.

The crystal growth was monitored by videotape in both Shuttle and ground-based experiments. A comparative analysis conducted post-flight assessed differences in color, density, hardness, morphology, refractive index, microscopic analysis, as well as electrical and thermal conductivity.

RESULTS

The videotape of the crystal growth process showed that crystals grown on Earth formed only on the upper half of the membrane: in contrast, the space-grown crystals formed symmetrically over the entire membrane. Moreover, the Earth crystals fell from the horizontal shelf formed at the midline of the membrane, whereas in the Shuttle experiment they did not.
All crystals were analyzed with Auger electron spectroscopy and scanning microscopy methods to assess relative carbon contamination in the samples. The relative amount of carbon contamination in the Earth-grown sample was larger than that on the space-grown sample.

CONCLUSIONS

An important factor in the science of crystal growth, is the concept of a "nucleation site". In short, a nucleation site is a site of impurity which helps to initiate growth of the crystal lattice. An analogous process occurs in liquids if there is a contaminant (such as a foreign particle) adhering to the container. Another example of a nucleation site is an eddy current produced by the Earth's gravity. There is a notable difference between the space-grown and Earth-grown nucleation process. The space-grown crystals do not need to be attached to the membrane or to other crystals as is the case with the Earth-grown crystals.
Richard Cavoli's experiment demonstrated that there is a great difference between the growth mechanism of crystals grown in space and those on Earth. Since more nucleation sites are available in the space-grown materials, crystallization occurred on all parts of the membrane as well as in solution.
The commercial applications of such findings include the production of finer, larger crystals for use in X-ray film, intensifying screens, and in preparation of better molecular sieves.

4.4 THE YOUNG ASTRONAUT PROGRAM

The Young Astronaut Program is a national education program designed to equip students at an early age with the interest and skills to live in the technological world of tomorrow. Space exploration and the U.S. Space Program are catalysts to excite the natural curiosity of children and to instill a sense of fun into mastering scientific and mathematical principals.
Since its creation in 1984, the Young Astronaut Program has become the largest space-related educational organization in the world. Over 500,000 young people from preschool to junior high participate in the specially-designed learning activities created by the Young Astronaut Council. Additionally, there are Young Astronaut members in 30 foreign countries.
Because the Program is sponsored by leading corporations, membership fees are far lower than the actual cost and value of the materials provided.

START A CHAPTER

The core of the Program is the school-based Young Astronaut Chapter. Original, high-quality, space-related curricular materials are distributed to over 25,000 Chapters throughout the year. Chapters consist of up to 30 members established in a school or community organization under a Chapter Leader. While the materials are designed for easy inclusion into daily lessons, some Chapters choose to function as clubs or after-school activities. Material is supplied to sustain up to three hours of activities per week. Actual implementation and time spent is at the option of the Chapter Leader.
The Chapter registration fee is $40 per year (children do not pay individually). In many instances, local business or community groups sponsor the Chapter fee.

SATELLITE MEMBERS

Individual students may join the Program as Satellite Young Astronauts. For $10 per year, they receive the Astronews Newsletter, Space Watch guide to monthly astronomical events, a membership card and certificate as well as the opportunity to participate in contests and other special mailings.

PRESCHOOL "YOUNGEST ASTRONAUTS"

The most recent innovative program initiated by the Young Astronaut Council is an exciting curriculum for preschoolers aged three to five introducing them to scientific and aerospace concepts. The program was developed in partnership with the U.S. Department of Health and Human Services and is being introduced to the 1,305 Head Start centers as well as to the nation's 229,000 licensed day-care facilities.

INTERNATIONAL ACTIVITIES

International student exchanges with the Soviet Young Cosmonauts and Japanese Young Astronauts have been extremely popular. Annual International Young Astronaut Conferences provide Young Astronauts the opportunity to meet with their counterparts from different countries. Pen pals exchange letters around the globe, broadening their horizons, as more countries launch their own Young Astronaut Programs. Affiliated programs currently exist in the former Soviet Union, Japan, South Korea, Republic of China and Canada.

CONTESTS

Chapters and individuals participate in art, essay, math and science competitions and contests. In addition to trophies and certificates, past prizes have included trips to shuttle launches, Japan and the Soviet Union.

CONFERENCES

National Conferences are held annually at exciting locations close to places of interest in aerospace.

FOR MORE INFORMATION:

Write to: Young Astronaut Council
P.O.Box 65432
Washington, D.C. 20036
The Young Astronaut Program is an educational organization established to encourage students in elementary and junior high schools to study science, mathematics, technology and related subjects. By receiving information and activities related to the Space Program through school chapters, young astronauts will gain the knowledge and skills they need for the coming decades. Students are encouraged to speak to their teachers about forming Young Astronauts chapters. Young Astronaut Pledge: "I pledge my best efforts to improve my grades in science, mathematics and related subjects, to learn about space and to help others towards these goals."

4.5 SPACE-RELATED ORGANIZATIONS AND PERIODICALS

As outlined in chapters 3.4 and 3.5 the space program is of vital importance for each and every one of us. And you can participate in it – even without being an aerospace professional or a GAS experimenter – by educating yourself and those around you about the space program, its current projects, and its long-term implications.
The major obstacles that prevent us from becoming a truly space-faring civilization and thus assuring our survival are not technical – they are political. But politicians are due for re-election – and if their voters are knowledgeable about space-related subjects, they themselves will have to favor a strong space program.
So tell those around you why space is so important! Start, for instance, by placing a space-related poster in your home, office or classroom, and talk to those who show interest in the subject. If you are a teacher, use your influence on tomorrow's voters to arouse their enthusiasm!

Eventually, join one of the many pro-space organizations which are open to the general public. These groups are of all sizes and special interests, from engineering to space colonization to planetary exploration. Of course just a few organizations can be mentioned here. If you contact them, they will be glad to provide additional information about their aims and activities:

NATIONAL SPACE SOCIETY

922 Pennsylvania Avenue S.E., Washington, D.C. 20078-1060
Founded in 1974 by the late Wernher von Braun, this is the oldest and largest of the societies that serve the general public, and membership in this organization is a must for every serious space enthusiast. Dues are $ 35.00 per year for regular members and $ 20.00 for those under 22, and include a subscription to AD ASTRA, an excellent full-color monthly magazine covering all aspects of space exploration.

THE PLANETARY SOCIETY

65 N. Catalina Avenue, Pasadena, CA 91106
Formed in 1980 by Carl Sagan with the objective of encouraging the exploration of the solar system and the search for extraterrestrial life. The society publishes a bimonthly magazine, "The Planetary Report", containing the latest information on these topics, and also offers a large variety of space related educational materials. Annual dues are $ 25.00.

FINAL FRONTIER

P.O.Box 534, Mt. Morris, IL 61054-7852
This bimonthly, full-color "Magazine of Space Exploration" provides exhilirating photography and in-depth reports. A two-year subscription is available for $ 21.95.

UNITED STATES SPACE FOUNDATION

P.O.Box 1838, Colorado Springs, CO 80901
The United States Space Foundation offers a variety of educational initiatives. Its mission is to enhance the education of our young people to prepare them for success in a space, computer and technology-oriented society. The foundation also sponsors week-long workshops for educators each summer. Its dues ($ 5.00 for Elementary School students, $ 10.00 for High School students, $ 20.00 for College students, $ 29.00 for teachers, and $ 35.00 for individuals) include a subscription to its monthly publication SPACEWATCH.

THE CHALLENGER CENTER FOR SPACE SCIENCE EDUCATION

1055 N. Fairfax Street, Suite 100, Alexandria, VA 22314
This organization was started after the Challenger accident. The goal of the Challenger Center is to promote space science education through the establishment of education centers in key cities in the United States. As of July 1991 nearly 100,000 students flew on simulated space missions at various Challenger Centers, and well over a million students and teachers participated in space-related teleconferences.

THE BRITISH INTERPLANETARY SOCIETY

27/29 South Lambeth Road, London SW8 1SZ, England
The objective of the BIS – which was formed more than 50 years ago – is to promote the advancement of space research and space technology. Dues include the subscription to the bimonthly magazine "SPACEFLIGHT". In addition the BIS publishes the monthly "Journal of the British Interplanetary Society".